CW01496508

Tasting Notes

the art and science of
pairing beer and music

PETE BROWN

Foreword by
Mark Radcliffe

For Liz, who fell in love with me amid bad beer and better mix tapes. And for the Green Man.

Published by the Campaign for Real Ale Ltd
230 Hatfield Road, St Albans, Hertfordshire AL1 4LW
camra.org.uk/books

© Campaign for Real Ale Ltd. 2025

The rights of Pete Brown to be identified as the author
of this Work have been asserted by him in accordance
with the Copyright, Designs and Patents Act 1988.

First published 2025

All rights reserved.
No part of this publication may be reproduced, stored in a retrieval
system or transmitted in any form or by any means – electronic,
mechanical, photocopying, recording or otherwise – without the
prior permission of the Campaign for Real Ale Ltd.

ISBN 978-1-85249-396-7

A CIP catalogue record for this book is available from the British Library

Printed and bound in the United Kingdom by Short Run Press, Exeter

Managing Editor: Alan Murphy
Design / Typography : Dale Tomlinson
Sales & Marketing: Madeleine Hardman
Cover Design: Malcolm Garrett

Every effort has been made to ensure the contents of this book are correct at the time of printing.
Nevertheless, the Publisher cannot be held responsible for any errors or omissions, or for changes
in the details given in this book, or for the consequences of any reliance on information provided
by the same. This does not affect your statutory rights. Some quotes in this book have been
lightly edited for readability as they have been transcribed from recorded interviews.

Track listing

About the author

Pete Brown is a British author, journalist, broadcaster and consultant specialising in food and drink. Across 13 books, his broad, fresh approach takes in social history, cultural commentary, travel writing, personal discovery and natural history, and his words are always delivered with the warmth and wit you'd expect from a great night down the pub. He is the regular beer columnist for the *Sunday Times Magazine* – the only regular broadsheet beer column in the UK. He was named British Beer Writer of the Year in 2009, 2012, 2016 and 2021, has won three Fortnum & Mason Food and Drink Awards, been shortlisted twice for the André Simon Awards, and in 2020 was named an 'Industry Legend' at the Imbibe Hospitality Awards.

Foreword Mark Radcliffe

Beer is rock and roll. It's as rock and roll as a scuffed leather biker jacket or a well-worn pair of jeans or aviator shades or a foot up on the monitors or a distorted electric guitar going from E to A.

I was a teenager in the seventies and the music was as exciting as the liquid offerings were anaemic. Glam rock, heavy metal, folk, prog, rhythm and blues, reggae and punk all thrilled whilst the Worthington E, Double Diamond and Red Barrel provided a flat, flavourless accompaniment.

If the music was the peak of artistic inspiration the beer was a feeble flow of flaccid frustration. Thank heavens then for CAMRA who, like The Beatles and The Stones and the Bowies and the Bolans and the Sabbaths and the Zeppelins and the Dylans and the Floyds and the Feelgoods and the Wailers and the Pistols and the Ramones, taught us that life didn't have to be that way. Blandness didn't have to be something we had to accept. We could dream of higher things.

Watching an incendiary live band is one of life's visceral pleasures. It's the gift that keeps on giving no matter how old you are. Watching an incendiary live band with a decent jar in your hand is even finer.

And it's not just rock music. It's orchestras and operas and brass bands and disco grooves and techno rhythms and hip-hop beats. They all go better with a beer.

These days the brews are of a far higher quality. Any decent live music venue will have a range of decent offerings available to enhance the range of decent acts they put on. Music is still as great as when I was a kid, and there are too many people to thank for that. But the ales are in a different stratosphere and we can thank CAMRA for that.

Here then are some suggestions of how best to enjoy a pint and a song. Raise your glass. The house lights have gone down. Here we go. 1, 2, 3, 4... Cheers!

How does a beer writer end up on stage at a music festival talking to 1200 people about neuroscience?

Q: Is this a joke, or is it serious? A: Yes.

Pairing beer with music? I know it *sounds* like a joke.

To be fair, it started off as one. And then, I met the professor.

I'd managed to get myself invited to a special 'sensory dining' dinner at the Edinburgh Science Festival, where we were told our perceptions of flavour were about to be challenged. Each course was designed to confound our senses and mess with our heads in a particular way. A panel of speakers sat on stage facing an audience of curious diners, and I was on stage with the scientists rather than in the audience, where I felt I belonged.

I'd somehow blagged the gig because I'd recently started doing events where I paired different styles of music with different beers. My wife, Liz, runs the Stoke Newington Literary Festival, and she had a spare slot in one of the venues, a room above a pub that seated about forty people. Did I have anything that I could use to fill the slot?

Sure I did.

This was 2010. Beer blogging was at its zenith, craft beer was starting to explode in the UK, even cask ale was showing signs of growth, and people had finally stopped laughing at me when I told them I wrote about beer for a living.

As part of this happy, hoppy renaissance, books and events promoting the idea of pairing beer with food were everywhere. This was part of a broader push to try to persuade people, for better or worse, that you could take beer just as seriously as wine. Predictably, wine snobs laughed at the idea. But even many ardent beer drinkers dismissed it, saying beer couldn't possibly work as a pairing for food.

If you think about it, this is ridiculous: every single cuisine and menu ever developed by humankind is based on one simple principle – that some flavours go together better on the palate than others. How could the many flavours in beer somehow be exempt from this otherwise universal rule?

At the time, I thought this resistance perhaps had more to do with snobbery: both direct, and inverted. Wine snobs will always perceive beer as inferior, and perhaps for longtime beer lovers, pairing beer with food just seemed too pretentious? Maybe it smacked of taking beer too seriously?

I thought: you think pairing beer and food is pretentious? I'll show you pretentious. So, I started pairing beer with music, for a giggle, to show that pairing beer with food was entirely sensible by comparison.

Why this particular combination? I was passionate about music long before I cared about beer. As a teenager, I wanted to be rock star, and then, after briefly trying my hand at singing, a music writer. Now, years later, I had my revenge: a wheeze in which I could wind up sceptical people even more than beer and food matching, mainly as an excuse to unite my own twin passions of music and beer, and selfishly inflict my taste in both on a curious, if not exactly willing, audience. I stood in front of a half-full room upstairs at the White Hart, plugged my iPhone into a speaker, and played eighties indie music at people while they tasted a flight of beers. I had my fun. And that's all that mattered.

Two years later, by the time I was onstage in Edinburgh, the idea had developed a little, including the breadth of the music I used. I was given just enough time to do my party piece, the climax of my evolving show. I had two pieces of music on my iPhone that represented a deliberate stretch from my usual white-boy-indie comfort zone: Claude Debussy's 'Clair de Lune', and Jimi Hendrix's take on Dylan's 'All Along the Watchtower'. Two beers were served to the hundred or so guests: Chimay Blue, the classic Belgian Trappist ale; and Beavertown's much-missed Smog Rocket, a stout made with smoked malt. Both beers were rich and dark, complex and multi-layered. But while the former is smooth and vinous with murky depths I almost feel I can swim in, the latter was challenging and astringent, slightly aggressive and in-your-face. (For more detail on this, see page 196).

I played the two songs and asked the audience to taste both beers with each one. Then I asked for a show of hands: which beer tasted best with which song? Instinctively – as I'm sure you will agree – it made sense to me that the smooth, classic beer would go with the swirling strings and harps that sound like they're being played by mermaids, while the bold, heavy bruiser would go with the squalling, dive-bombing guitars.

The room was evenly split. My experiment had failed. So I was surprised afterwards when one of my fellow panellists – whose presentation had worked far better than mine – approached me. 'Do you realise that what you just did is very close to experiments we're conducting at Oxford University?' he said.

'No way!' I replied, a lot more politely than the immediate, instinctive response in my head.

Professor Charles Spence is Professor of Experimental Psychology and Head of the Crossmodal Research Laboratory in the Department of Experimental Psychology at Oxford University. He explained that this particular branch of not-quite-neuroscience (he often uses the term 'neuro-gastronomy') explores how our different senses affect each other – how colour or shape affect our perception of sweetness, or how a fair chunk of what we 'hear' is profoundly influenced by what we can see. Charles and academic colleagues around the world had been mapping different sonic pitch, and even different musical styles, with the basic tastes of sweet, sour, salty and bitter, and finding pretty strong correlations using rigorously controlled scientific experiments that had been written up in peer-reviewed publications. I'd accidentally stumbled across something serious – even if my particular version hadn't worked this time.

Or so I thought.

We'd been chatting for a few minutes, when a member of the audience sidled up to us. 'That was so clever how you used the music to change the flavour of the beer!' he said to me.

For the second time in a few minutes, I edited a crude outburst of disbelief into something appropriate for the occasion.

'Pardon?'

'Yeah! I loved how when the classical music was on, the Chimay tasted gorgeous and the Smog Rocket was horrible and bitter. But then when Hendrix came on the Chimay was watery and bland and the Smog Rocket tasted just right. Really clever!'

This was news to me. But then, it has often been observed that some of the greatest discoveries in science are accidental. Alexander Fleming discovered penicillin because he didn't clean his equipment properly. Kodak engineer Harry Coover was trying to make clear plastic gun sights when he accidentally created superglue. Now, I had discovered how to change flavour simply by changing the sounds that go with it. Does this mean I am destined to join the ranks of Fleming, Pasteur and Hawking as one of the great scientific minds of the age? History must be the judge.

On the road

After that, things got serious quite quickly. I started to read some of the many papers Charles Spence and others had written in academic journals. Charles actively encouraged me to dig deeper, to think more.

In 2014, I was invited to do my party piece at Bristol Food Connections, a new food and drink festival. Dan Saladino, producer of BBC Radio 4's *Food Programme*, was involved in the programming, and set up the gig. About three weeks before the event, Dan said casually, 'Oh, by the way, Pete, Lord Hall, the Director General of the BBC, is probably coming to your beer and music thing.'

I froze. Standing up with an iPhone in one hand and a beer in the other was fine for a bunch of punters in a room above a pub who were, let's be honest, only there for the beer. But it really wouldn't do for the DG of the BBC. I mean, if I impressed him, he might commission me to do more live shows. A radio programme. A TV show! Obviously, none of this was ever going to happen. But you can't succeed in writing books if you don't cultivate utterly deluded ideas about how rich and famous your work is going to make you.

You'd never stay the course on all that lonely work if your hopes and dreams had any contact with reality.

So, full of ideas about being spotted as the host of a new primetime BBC1 series about beer, I plunged into the science. I read academic papers and learned about brain scans and crossmodal correspondences. I also learned how to download and edit videos from YouTube. How to mix fragments of different songs. I wrote more than two original jokes. And I squeezed the whole thing into an all-singing, all-dancing[1] PowerPoint presentation. Back in the room above the White Hart, I previewed my new multi-media extravaganza in front of a hand-picked group of friends. Apart from the one who walked out because he thought the new joke about jazz[2] was taking the piss out of him personally, everyone seemed to enjoy it.

I wrote a script that I learned off by heart, so I could focus on stealing the comedian Dave Gorman's use of PowerPoint slide transitions for comic timing. I stole a brilliant experiment using Skittles from one of the Edinburgh Science Festival presenters. I stole quite a lot of Stewart Lee's microphone technique. I stole widely and I stole well.

Apparently, the DG was in the back of the room for almost five minutes. He never called. He never wrote. But I now had a real stage show because of him. Richard Thomas *did* call, though.

Richard is a friend who, at the time, was programming the spoken word stage at the Green Man Festival in the Brecon Beacons. He's had a long and storied career in the music industry. And Green Man, unlike any other mainstream music festival, has a beer tent with about sixty locally-sourced cask ales and another thirty or so local real ciders. My challenge was simple: do my beer and music show using the beers and bands playing the festival.

The first year Richard booked me, I was on at 3pm on the Sunday afternoon. If you've ever been to a weekend-long music festival, you'll appreciate that this is the time when everyone is filthy, tired, hungover, drunk and euphoric all at once.

We had to send a runner across the field to the beer tent when everyone realised I actually needed the beers there to taste.

Green Man is a thirsty festival, and by this time, most of the cask beer had gone. A crew member was carrying jugs of Wrexham Lager and Green Man Growler, the only beers they had left, spilling most of it on the mud plain between the tents. Twenty red-eyed revellers gazed balefully from a sea of a thousand empty chairs. They drank their free beer, and left.

Incredibly, Richard booked me again the following year. What if we did it on the Friday morning, the first event in the spoken word tent (which no one went to anyway) when we could take our pick of the freshly racked beers and talk about bands that hadn't actually played and gone home yet, so people could go and see them afterwards, perhaps with a pint of the beer we'd used instead of a tiny sample?

The thirty people who turned up that year didn't seem quite so miserable. The following year, I realised the spoken word tent had a screen, and I could take my slideshow along. Sixty people came, and enjoyed video previews of my pick of the bands that were about to play, alongside bright, foamy cask ales.

The fourth year, I turned up an hour early to check my tech, and there were about a hundred people sitting in the front row.

'We don't start till twelve,' I said.

'Yeah, we know,' said a mean-looking bloke in the middle. 'We got here late last year and missed the free beer.'

I went backstage to prepare. When I walked out in stage an hour later, the tent was full. Liz and about five other crew members spent an hour running up and down the aisles trying to serve four differ-ent beer samples to over a thousand people, and in doing so, became Green Man legends in their own right. They were stopped and congratulated by happy festival-goers for the rest of the weekend.

I've done every Green Man since. Richard no longer programmes the line-up, and every year I assume it will be my last. But they keep inviting me back. About two-thirds of the audience consists of people who come along every year, and chant the punchlines of over two jokes with me, like the choruses of the songs they'll be singing along to later. They know that I dig deep into the line-up to look for acts that I've never heard before, making sure my show is a mix of styles, genres and cultures, and they use me as a primer for what to listen to and what to drink over the weekend.

In 2023, I showed a beer called Afghan Pale Ale from the Grey Trees Brewery in Aberdare.[3] I made the mistake of saying that this was probably my favourite beer at the festival. About an hour later, when I was finally free after my event, I went to get a pint of it because I hadn't managed to get a taste during the show itself. The beer tent was arranged by brewery, and it looked like a grinning mouth with a tooth knocked out. In a wall of gleaming casks, the Grey Trees area was smashed. Their whole stock for the weekend had gone by 2.30pm on the first day. The staff were standing there looking glazed and slightly traumatised, not entirely sure what had just happened. I never did get a pint. But I realised that the show really was having an impact.

There's a bookshop tent next to the main stage, and every year they get whatever stock they can find of mine that's currently in print, and I sign a few books. And every year, 6 Music Dads and their long-suffering wives and children ask me when the book of the show is coming out. Finally, fifteen years after I first came up with an idea that was a piss-take of beer and food matching, here it is.

Pairing art and science

There is more than one layer, more than one reason or angle, to pairing a beer with a song or piece of music. I must have done my event over a hundred times now. At one gig, a table at the back revealed that they were all neuroscientists. I spoke to them afterwards, saying I felt I should apologise. They said no, the science bit, such as it was, had all been correct.

I did another gig in a private dining room at a Michelin-starred restaurant in the heart of London. Professor Charles Spence came, to see how I was getting along. Just as he had done years previously at the Edinburgh Science Festival, he came up to me afterwards and began, 'You do realise … that by going so far into the science and trying to get it right, you've lost all the fun and humour that made it so appealing the first time?'

With an Oxford University professor as my spirit guide, today my live show boasts cutting edge neuroscience *as well as* almost a dozen jokes. OK, maybe six or seven good ones. But those first three still knock it out of the park.

The point of me telling you all this, before we dive into the book, is to try to sell you on the idea that you can enjoy this conceit of beer and music pairing in any number of ways. Yes, it's a joke. And yes, it's very serious. It's about taking two different cultural products, mashing them together and having fun with them. It's also about going, hang on, they go together in more ways than perhaps we thought. If you're into one, you're probably into the other. Both beer and music are about enjoying the world around us. Sometimes on our own, but especially with other people.

If you want, you can go further than that. Modern neuroscience is in its infancy. Here, in the third decade of the 21st century, we know more about the birth of the universe than we do about how our brains sort and interpret information from the immediate environment around us.

Doing my shows, I learned quickly that the natural resting face of someone who is thinking 'This is a load of rubbish,' is very similar to that of someone who is concentrating really, really hard and having their mind opened. I suspect that I've had more than a few audience members who started at the first position and moved to the second without needing to shift their frown by a millimetre.

If you want to jump straight to the pairings and flip through to find your favourites, be my guest. This is your copy of this book, and you can do what you like with it. All that we perceive is subjective, our impressions influenced by our individual lived experiences and the preferences they shape.

But if the idea of having your mind blown, if the concept of being lifted out of your established groove and challenging your perception of reality gives you a flutter of excitement and intrigue – well, from here on in, with a glass in your hand, just turn one page at a time. Switch on your mind, relax and float downstream...

SIDE ONE

The Theory

Your brain is lying to you.

You have more than five senses.

And sour beers have a different sonic pitch from IPAs.

How flavour works – and why it's a lot more complicated and uncertain than you probably thought

Putting words in your mouth

If you're reading anything about food and drink – a recipe for beef rendang; a list of the ten best beers in Bristol; an article about how beer is brewed – it would not be unreasonable of you to expect to read a description of how it tastes.

Ideally, that description should be accurate. It should make sense to you. And unless someone is doing a hatchet job – such as, say, a detailed and accurate analysis of how Corona lager really tastes when you get rid of the lime wedge and pour it into a glass at the recommended serving temperature for lager – it should be appealing. It should make you want to put it in your mouth.

When I tell someone that I write about beer, they automatically assume that this is what I'm talking about. But I *hate* writing tasting notes. It's the most difficult and frustrating part of the job. I'd rather write about beer from almost any other angle than how the stuff actually tastes once it's in your mouth. I can do it pretty well. It's just really hard.

But you knew that already. Because you find it difficult too.

I was in the pub with a mate who loves beer, but isn't a geek. He appreciates the difference between cask and keg, for example, but would never describe himself as an expert. When we go for a beer together, he asks me what's good, and usually enjoys my recommendations.

We were sitting at a table with two pints of London Pride before us, served in dimpled jugs.

'Can you describe what that looks like?' I asked him. 'In as much detail as you can.'

'Sure. It's glass, so it's transparent, and you can see the beer inside. The glass is roughly cylindrical, with a curved handle on the side. It's about eight inches high, about four inches across. The dimples on the side of the glass refract the light a bit, so the colour of the beer is broken up with bits where you can just see the silveriness of the glass. The beer is mid-brown, amber with a hint of red. It's clear – you can see through it – and the light from the bar makes it glow a bit which makes it look appetising. There's a thin head of white bubbles, and some gentle carbonation is coming up through the beer, but the head is slowly dissipating as the bubbles burst.'

'Thanks!' I said. 'Now can you tell me what it tastes like?'

He took a sip, and frowned.

'Um… it tastes like beer?'

'What does that mean?'

He started to look flustered. 'Er… it's bitter. But not in a bad way. It's sort of… rounded? Balanced, I suppose. Grainy? A bit biscuity?'

That was it. He loved the beer. But he was visibly uncomfortable trying to describe its flavour to me in any meaningful way.

Neither my mate nor anyone else needs specialist training of any kind to describe how a pint of beer looks. The description he gave here hopefully creates a very clear image of that pint of Pride in your mind. But after twenty years of sensory training, beer judging and evaluation, even I wouldn't be able to describe the flavour of the beer in anything like the same level of detail in which I can describe its appearance.

There are various ways we try to solve this difficulty, depending on who we are. None of them work as well as they might.

In quality control and in competitions, the language that brewers, judges and professional tasters[4] use around flavour is very specialised: this beer is quite estery; that wine has ethyl acetate; this cider is acetic. These are precise technical terms, very useful for evaluating quality and potential flaws, but useless for communicating the character or appeal of flavour to a non-expert. And yet, sometimes, you still see some of these words on labels, as if they'll make people think 'Hey, complex yeasty esters! I hope there's just a hint of diacetyl in there somewhere too. Let me get this bad boy in my basket!'

Even now, I'll occasionally fall into the trap of describing a beer as 'hoppy' or 'malty'. These words are not quite as bad as the technical terms above, but a good writer should still avoid them if possible. I worked in beer for two years with people telling me a particular beer was hoppy before I had any understanding of what 'hoppy' was. 'I'm sure it is,' I would say. 'I'm not doubting you. But can you describe what "hoppy" means? What is it that I'm tasting right now that comes from the hops?' No one could tell me. Today, more people could. You might be referring to the bitter, drying finish, or you might be referring to fruity, grassy or spicy aromas. So why not just say that instead?

At the other end of the scale, there are words like 'delicious,' 'yummy', and, if you don't mind your friends secretly hating you, 'nom'. But these words do nothing to describe or evoke flavour; they merely tell you that the person saying them is enjoying the flavours they're experiencing. They could be eating a burger or a piece of tuna sashimi, drinking a pint of Timothy Taylor Landlord or a glass of Picpoul for all we know.

In 1982, BBC2 launched a new programme, *Food and Drink*, which introduced wine expert Jilly Goolden as a regular reviewer. She rejected wine's version of the technical language above, and attacked the issue of flavour description head on, taking no prisoners. In her mouth, wine could taste of pear drops, liquorice, or even rubber. It could be 'creamy with a touch of banana,' reminiscent of compost or rotting cabbage (in a good way). She described one memorable wine as being like 'a wooden bra'.

She was ridiculed and adored in equal measure. She became a household name, was endlessly parodied, even ending up with her own *Spitting Image* puppet. But she is responsible for reinventing how we talk about flavour. 'I do [it] in adjectival overdrive to give viewers at home, who haven't got a glass of it in their hand, a chance of envisaging what it's like,' she said years later in a newspaper interview. 'I examine the wine in huge detail on nose and palate, and the scent and flavour recall everyday scents and flavours that

I attempt to build up into a "taste picture". In that way I attempt to describe the elusive bouquet and palate of the wine.[5]

This approach is what we who must write tasting notes now all follow, to greater or lesser degrees. When we describe Mosaic hops or Kiwi Sauvignon Blanc as having a 'cat pee' character to it, we know it's not a perfect simile, but once you get used to this approach, you're going to know what we mean, and won't be repelled by it as a description. Hopefully.

It still has severe limitations though. I might describe a best bitter or mature cheddar as 'nutty'. Okay, fine, so what do nuts taste like? Well… nutty. At a push, I could say 'nutty' is earthy, slightly woody, a little sweet. But hang on – what do earth and wood taste like? I've never eaten either. And if my pint of best tastes nutty, and nuts taste earthy, does that mean my beer tastes earthy, too? Cumin is often described as having an earthy flavour. So how come my beer tastes nothing like cumin if both the beer and the spice taste earthy?

This is the best we can do when trying to describe flavour to other people. But linking flavours in chains of similarity like this relies on us sharing a common understanding of what certain things taste like. Most of the time, we do. If I say this beer is nutty, you and I probably have a similar idea of what nuts taste like. And if we've both tasted enough beers, even if 'nutty' isn't quite right, you know what quality in the beer I'm getting at.

But we don't always appreciate that some terms mean different things to different people. I chose 'nutty' as an example because Liz has a potentially fatal allergy to nuts, and will never experience their flavour. How can I communicate it to her?

Let's take another example.

For years, I described the intense, fresh aromas of certain North American hop varieties as 'floral'. To me, and many of my peers, this word perfectly conveyed how these aromas were different from the spicy, loamy character of British hops. But I stopped doing this after observing my words used in a market research focus group of middle-aged beer-drinking men in Newcastle, who spluttered and exclaimed, 'Floral? No thanks! I don't want bloody *flowers* in my beer!'

Maybe these difficulties are the reasons why an award-winning copywriter at the first advertising agency I worked at had a rule: never, ever, try to write an ad that describes flavour.[6]

In 2024, I was visiting the Cruzcampo pilot brewery in Seville. The brewmaster there was a young woman who was doing a brilliant job of describing the flavours of her beers in a tutored tasting, in a language that was not her own. At one point she said, 'To describe flavour, we must have a memory of it,' which perfectly and poetically sums up the whole conundrum. And raises the obvious problem: what if my memories of flavour are different from yours? Which of course, they are.

If we can't talk about flavour properly – if we can't describe it – then we can't really understand what it is, or how it works. And that's what's behind the whole idea that inspired this book. We eat and drink every day. But we don't really know how our appreciation of food and drink works. That means it's open to misunderstanding, bafflement, trickery and deception. And, if you're up for it, a very, very tasty journey of discovery.

But before we set off on that journey, it's worth asking – why are we in this situation in the first place?

The 'lower order senses'

Just as we can describe what a pint of beer looks like pretty well, when we hear a piece of music, we can say whether it is harmonious or discordant, quiet or loud. We don't have to know too much about music to identify what instruments are playing, or (mostly) whether voices are male or female. We can tell what direction the music is hitting us from, make a good guess about how far away the source of the music is, and even tell what kind of space it's playing in. You might even have an 'earworm', where your recall of a piece of music is so perfect that your brain can still 'hear' it playing on a continuous loop. We have no comparable sense memory for flavour.

So why is our understanding of flavour so under-developed compared to our appreciation of what we can see or hear? A big part of it rests in our philosophical construct of how we interact

with the world. It's a belief system that goes back to Aristotle and
Plato. It was Aristotle who first defined the five senses we still think
of today: sight, hearing, touch, taste and smell.

We actually have many more senses than this. There's thermo-
ception – our sense of temperature, which is perceived separately
from our sense of touch. Magnetiception, our 'sense of direction',
works in relation to the earth's magnetic fields. Not every human
has it to a useful degree, but it's a real thing, and is a much more
important sense for many birds and insects. The 'proprioceptive
senses' include our perception of the position and movement of our
limbs and body – you know where your arms are in relation to your
head without looking or feeling for them. There's also the sense of
effort, the sense of force, and the sense of heaviness. You know, for
example, whether the pint glass in your hand is empty or full. Your
sense of touch tells you that the cool, smooth glass is in your hand,
but not how heavy it is. The receptors involved in proprioception
are located in the skin, muscles, and joints. We can recognise all
this when it's pointed out, even as we talk about the mystical 'sixth
sense', which no one really disputes, but no one has really defined.
So why do we persist in talking about only having five senses?

Some philosophers argue that this is because the traditional five
are the ones through which we directly perceive the outside world,
whereas the other senses are more internally focused. (I'd dispute
this, but that's a different bar-stool conversation.)

The classical five senses are traditionally arranged in a hierarchy,
a philosophical league table of merit. At the top, still crushing it
after thousands of years, is sight. Plato spoke of 'the eye of the soul',
and compared the eye to the sun. At the beginning of the *Metaphysics*,
Aristotle wrote, 'We prefer sight, generally speaking, to all other
senses […] Of all the senses, sight best helps us to know things.'
We might have a *vision*, prompted by *insight*. To think is to *reflect*
on things, and to explain is to *illustrate*.[7] We see (*see?*) the world in
visual terms.

Hearing, like sight, has for the last two thousand years or so,
been seen – or rather, *regarded* – as a 'higher', more spiritual sense.

In Renaissance England, music, and thereby hearing, were linked to order and disorder. Order and harmony in music meant order and harmony in the universe, and in ourselves. Martin Luther also observed that you need hearing in order to be able to hear the Word of God.[8] Sight and hearing have, for most of civilisation, been linked by the leading thinkers of the day with the higher human capabilities of ethical thought and higher reasoning, not to mention the creation and appreciation of music, literature, drama and art.

Socrates believed that there was a division between body and soul, and that the body played no part in the attainment of worthwhile knowledge. Touch, taste and smell belong to the body. They involve feeling things and putting them in our mouths. As such, they are linked to appetites. Appetites are all about bodily functions, a long way from the delights of the mind that the philosophers and preachers were so fond of. At worst, our appetites can lead to lust, one of the seven deadly sins. For people like Socrates, food, drink and sex got in the way of true enlightenment.

These attitudes persist even today. The pursuit of sensory pleasure for its own sake is the true definition of *hedonism*, a word still couched heavily in disapproval. But sex, eating and drinking are pleasurable for a very important reason: if they weren't, we might not do them. And if we didn't do them, lofty minded Greek philosophers would never have been born, because humanity would be extinct. You can't seek enlightenment if you're dead. Socrates may have been one of the cleverest men who ever lived, but he was also an idiot.

For Aristotle, touch was the most problematic sense. For him, it was rooted at the foot of the table. Not only is touch part of the animal, physical, world; *it's all over the body! Ugh!*

Later philosophers gave the sense of touch a critical reappraisal. It may be all over the body, including the yukky bits and the naughty bits, but our sense of it is focused mainly on the hands, which we use to paint, sculpt, write and play musical instruments, creating the lovely, lofty things we see and hear. So it can't be that bad. The elevation of touch left taste and smell propping up the league.

Incredibly, this hierarchy hasn't really been challenged since.

According to international research in 2018, the weight given to the top three varies across cultures. But taste, followed by smell, always come bottom in terms of how comfortable humans feel thinking about them.[9]

This snobbery around the senses has been compounded by circumstance. For most of history, food and drink were scarce, and the average person didn't really get the opportunity to be a gastronome. The main smells in any city would be body odour, woodsmoke, and shit. Which is why poor old smell has an even worse rap than taste. We now have sewers, deodorant and toothpaste, and breathing in through your nose is no longer as fraught with risk as it was when Immanuel Kant argued that the 'dispensable' sense of smell 'does not pay us to cultivate it or refine it in order to gain enjoyment' because it detects more that is foul than is pleasant.[10]

If something is 'tasty', we want to try it. If something is 'smelly', we don't want to go anywhere near it. We disguise smell with words like 'odour' – still not great – 'aroma', or even 'bouquet' – because anything in French is classier than its English equivalent. Even Jilly Goulden baulked at talking about smell, instead using the word 'scent', a term synonymous with perfume.

So, smell ends up even below taste, right at the bottom of the table. This is ironic, because it some ways it is the most evocative sense of all.

Getting a grip on taste, smell and flavour

Taste and smell are never seen out together without a third concept that is not one of our main senses, but often gets talked about as if it is: flavour. The relationship status between these three is … well, it's complicated.

We often use 'flavour' and 'taste' interchangeably, as if they mean the same thing. The *Collins Thesaurus* gives flavour as a synonym for taste, and taste as a synonym for flavour. When we use these words as nouns to describe something, they're simply two different but similar words to describe the properties of food and drink.

Yet we also use both words as verbs, and when we do that, they mean quite different things.

If we're *flavouring* something, we're adding some adjunct to it to make it 'taste' different. In beer judging, the 'flavoured beer' category consists of beers that have had fruit, herbs or spices added to them – flavours that you don't get in conventional beer. Google 'flavoured foods', and you discover an entire industry devoted to artificially creating fake flavours for ultra-processed foods.

But when we're *tasting* something, what we mean is we're eating or drinking it carefully and mindfully so we can experience everything it has to offer, as opposed to just getting it down us. So while we might go to a beer *tasting,* we don't go to a beer *flavouring.* The ironic thing about this is at a beer tasting, the most important thing you can do is smell. When a waiter asks you if you would like to taste the wine you've ordered in a restaurant before pouring, if you want to show that you know what you're doing, you don't taste it at all – smelling it is enough to know whether it's corked or (more likely) oxidised.

Still, we persist in using 'taste' and 'flavour' interchangeably, and by doing so, we're confusing ourselves before we've even started. Let's imagine we're at a formal beer tasting. Here, amid the elegant sample glasses, starched white table cloths and beer writers whose notebooks are getting slightly messier with every page, we can establish a base camp. From there, we can dive in a bit deeper and show why some of what even the self-appointed experts say is incomplete. (I'm not being smug about this. The below is almost verbatim of how I used to do this, not a piss-take of anyone else.)

What the tasters say

'Okay, so the first thing we do is look at the glass. The colour gives us some big clues as to what to expect. Next, we give it a big swirl. That's it, just like you see the wine critics do it on the telly! This releases the volatile aroma compounds in the beer that give it its aroma. That's why it's important to then take a big sniff from the glass. Because when we talk about flavour, about eighty per cent of what contributes to that is actually the aroma. We only have five basic tastes on the tongue, but there are

millions of different aromas. That's why you can't taste anything if
you drink straight from the bottle – and that's why Corona insists you
should drink straight from the bottle! So, swirl, hold the glass under your
nose and give it a good sniff. This is your first big clue to the character
of the beer. What are you getting? Citrus? Yes, good. Caramel? Great!
Drains? Ooookaaaayyyy … Mandarin orange pith? Oooh, get you!
Now take a big glug. Wash it all around your mouth to make sure it gets
all over your tongue … and swallow! Because the big difference between
beer tasting and when you see wine tasting on the telly is that we don't
spit, we swallow! That way you get the full character of the beer.'

This guided tasting technique is going to give you a far deeper
appreciation of the beer than if you just swigged it while chatting
with friends or watching TV. But going even deeper allows us to
get even more – and also starts to explain why your perception of
flavour can easily be fooled, and how the basis of this is the fact that
our different senses are not independent from each other, but in
fact overlap and sometimes change our perception of what each
one is trying to tell us individually.

What is flavour? Really?

Flavour is the what you get when you combine aroma and taste.
    Actually, it's a bit more than that. The official International
Standards Organization (ISO) definition of flavour is a 'Complex
combination of the olfactory, gustatory and trigeminal sensations
perceived during tasting.'[11] Taste is a part of flavour, not the same
thing. But there's even more going on than just taste and aroma.
Let's take it step by step. Imagine – if you can – a glass of beer in
front of you, with enough room in the glass so you can swirl it
without spilling it. Or even better, go and pour yourself one.
Make sure it's a good one, with some depth and character to it.
    Revisiting what a typical taste tutor would say, we'll come
back to *looking* at the beer later. It's an important step, but it's an
additional one. Let's cover the basics first.

Aroma #1: Orthonasal aroma

First, swirl and sniff your beer. We do this to get a better experience of its aroma. Aromas consist of volatile (i.e. gaseous) chemical compounds. On the slightest whim – a shake, a squeeze, a swirl – they break free from the glass (in this case) and waft through the air. As soon as they're free, they dissipate, so sometimes we can detect the merest hint of an aroma that grabs our attention, teases us, then disappears. That's why we sniff as hard as we can, to try to draw them in.

There are literally billions of aroma compounds out there in the world. Many of them derive from essential oils in plants – and this is why hops are so special. Over 400 different essential oils have been identified in hops. They're all found in other plants, but only the hop has all 400 of them. That's 400 different aromas to start with, before brewers even start to combine them. And the remarkable thing is, your smelling equipment is theoretically capable of detecting them all.

Take that deep sniff again. The volatile aroma compounds escaping from the beer are sucked into your nasal cavity, which has evolved to be large enough to give them room to float around. If they (and we) are lucky, they get picked up by the olfactory bulb. This sits on the roof of the nasal cavity. In diagrams, it's represented as a terrifying-looking structure that looks like some sort of weird tree root growing down into the nasal cavity from the bottom of the brain.

This extraordinarily highly tuned instrument feeds directly into several key parts of the brain: the amygdala, which processes memory, decision-making and emotional responses; the hippocampus, which is crucial to short- and long-term memory, spatial navigation and conflict avoidance; and the orbitofrontal cortex, which is thought to be vital in calibrating emotion and reward in decision making. Our perception of aroma shoots deep into memory, instinct and emotion before we can even think about it, which is why smells can sometimes be so evocative on such a deep level.

One early estimate suggested we are capable of distinguishing 10,000 different aromas. But recent experiments testing people with tiny nuances in aroma suggest that, if we had an infinite

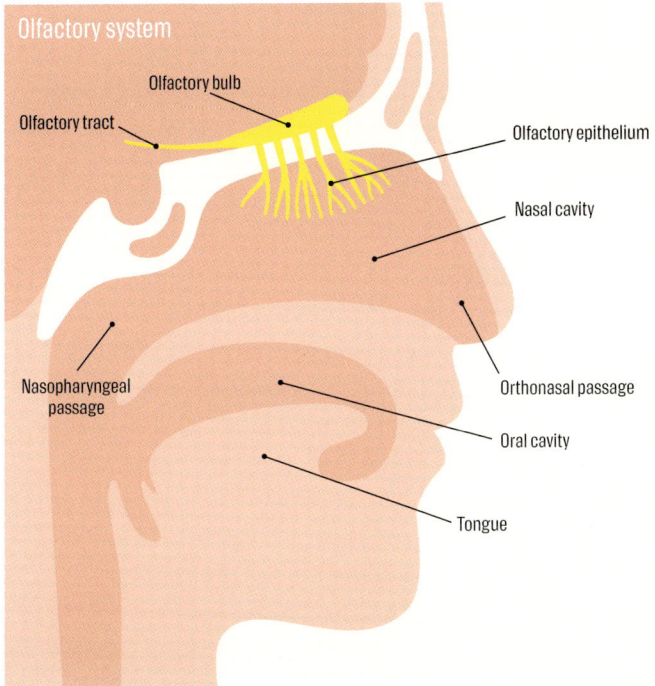

Olfactory system

Olfactory bulb

Olfactory tract

Olfactory epithelium

Nasal cavity

Nasopharyngeal passage

Orthonasal passage

Oral cavity

Tongue

amount of time to run those experiments against all the aromas, we are capable of distinguishing 10 *trillion* different nuances.

We refer to the action of taking in a deep draught of aroma as 'sniffing' something. And yet, going back to our restaurant wine tasting, if the sommelier asked the customer if they would like to 'sniff' their expensive bottle of wine, this would be seen as condescending. If a person is 'sniffy', we mean they are over-critical, probably in a way they have no right to be. It almost goes without saying by now that, in comparison, if we describe a person as 'tasty' we are admiring them in a sexual or physical way – and we're back to appetites again.

When a 'smell' is bad, it really stinks. When 'aroma' or 'scent' is good, it's the most romantic of all the senses. That fleeting nature, that deep, searing evocation of sense memories that often escape your attempts to pin them down, is the whole basis of a global perfume industry worth around $40 billion annually. In food and drink, aroma is an early warning sign, a flirtation, an overture, a secret weapon with which to impress our peers. It's dangerous and primal, common and elitist, mystifying and everyday, universal and almost universally misunderstood, the most underrated and cruelly abused of the commonly defined five senses.

Taste

'Taste', as the word suggests, is the effect of direct stimulation of our taste buds, which are in our mouths. See? Not all of it is complicated.

Tastes are created by more chemical compounds, which behave differently from aromas. While aroma compounds are released as particles into the air, taste compounds must be soluble in water or saliva to be able to reach our taste buds. These are located mainly on the upper surface and sides of the tongue. But they're also on the soft palate (the roof of the mouth toward the back of the oral cavity), and pharynx (the bit between the mouth and the throat).

Taste on its own is quite crude: we can detect a mere five tastes – or maybe six, seven, or eight depending on who you read. These are sweetness, sourness, saltiness, bitterness and umami. And, arguably, possibly, fattiness, metal and carbonation. Over time, different candidates for what qualifies as a taste emerge and are argued over. Current discussion revolves around fattiness (or is just texture?) and metallic (or is it actually more of an aroma?) The most recent successful addition to the canon was umami, first coined in Japan over a century ago, and now just about accepted globally.

Umami translates literally as 'savoury deliciousness'. In 1909, Japanese physical chemist Dr Kikunae Ikeda proposed that 'While [umai, or delicious taste] is based on a subjective sensation [...] there can be little doubt that another taste exists [...] I propose to call this taste 'umami'. He then proved his point by isolating the sodium salt of glutamic acid, patenting his production method, and inventing monosodium glutamate (MSG), still the perfect example of 'savoury deliciousness'.

Take a generous swig of your beer now (what do you mean, you already did?) Swirl it around slowly, coating your whole mouth. For a long time, we thought the receptors that picked up these different tastes were on different areas of the tongue. Some food and drinks books (including some of mine) still claim that we perceive sweetness at the front, sourness and bitterness down the sides, umami in the centre and bitterness at the back. It's now been conclusively proven that we get all tastes on all parts of the tongue, though some researchers maintain there are areas of concentration, with heavy overlaps. The fact that science is still not entirely certain about this is a perfect illustration of where we are with our understanding of flavour.

Aroma #2: Retronasal aroma

Hang on a minute. So, we're saying that most of what we perceive of as flavour is aroma. And we experience aroma in the nasal cavity. But here your are, savouring your beer with your mouth closed, because you were well brought up, and you're still getting all the flavour sensations I said we only get from smell. How come?

Thanks for paying attention, I'm glad you asked. The answer to this question doesn't just help us understand flavour properly; it's absolutely solid proof that our senses interact and overlap, and our brains sometimes misinterpret what the sensory organs are telling them.

When we 'smell', we automatically think this means detecting an aroma that's outside the body. But we 'smell' from inside our mouths, too.

That's not as gross as it sounds. Once food or drink is in the mouth, the action of chewing or sloshing, plus the internal temperature of the mouth (37°C) release more aroma compounds. With our mouths closed, we have to breathe out through our noses. When we do, our out-breaths carry these additional, newly released aroma compounds up the nasopharyngeal passage at the back of the mouth, and past our olfactory bulb once more. This is called retronasal olfaction, and humans get a lot more aroma impressions from this than we do good old-fashioned orthonasal olfaction (sniffing.) This is why flavour perception in the mouth can be quite different sometimes from what we smell. With a piece of stinky cheese, the ammonia compounds are quite volatile and assault your nose. In the mouth, as well as taste and texture (more on that below) softer, subtler aroma compounds are released and you get them retronasally.

But we never think of this as aroma. Because the food or drink is in our mouths, and that's where the chewing or swirling is going on, the brain gets it wrong and perceives retronasal olfaction as being experienced in the mouth, when in fact it's happening in the nasal cavity. So we think of it as taste rather than aroma. Maybe you don't need to feel quite so bad about getting taste and flavour mixed up. Your brain is just as confused as you are.

This might all sound far-fetched. But it's the reason why you think you can't 'taste' anything when you have a head cold. You can in fact still pick up the basic tastes just fine – you can still perceive whether something is salty or bitter – but because you can't breathe out through your nose, and your olfactory bulb is covered in snot, you're not getting any retronasal aroma. If you don't believe me, carry out this simple experiment below.

Experiment: Skittles

Take a packet of Skittles – or any other strongly flavoured sweets with mixed flavours. Once you have the packet open, with one hand (or a clothes peg) pinch your nose shut. With your other hand, pick up a Skittle while making sure you don't see which colour/flavour you've chosen. Put in in your mouth and chew.

What can you taste?

Now, keeping your mouth closed, take away the hand/peg that's blocking your nose, and breathe out.

What can you taste now?

Trigeminal effects

Apart from taste and smell, some scientists argue that another aspect of our experience of flavour, which never gets enough credit, is texture. The mouth contains nerve endings and other receptors as well as taste buds, and they all add to the sensations we get when we're eating or drinking. This involves, but is not limited to, the sense of touch, so these various sensations are grouped under the heading 'trigeminal'.

The most interesting example of this is the 'heat' of chilli. The *taste* of a chilli pepper is actually quite sweet. The spiciness of it is not a taste or aroma. Neither is it a thermal sensation – a chilli can be cool to the touch before you pop it into your mouth, and yet you'll still say it 'burns'. The 'heat', which is not heat at all, can even make you sweat and go red in the face. It's actually a chemically induced irritation which is best classified as pain, though pain is something we experience through our sense of touch. This is not quite the same thing.[12] This is why that that 'trigeminal' word does a lot of heavy lifting – it's easier to say what it is not, than what it is.

The 'coolness' of peppermint is both a counterpoint and another example of the same sensation. Acidity might be a taste, but at a high enough intensity it also gives a trigeminal 'bite'. Astringency is an interesting one. Once thought to be a taste, it's now regarded as a trigeminal effect, created by tannins in tea, cider apples, red wine or wood-aged beers. Tannins act on the proteins in saliva to dry out the mouth. It's a highly subjective experience: I love bone-dry ciders and Fino sherry; Liz really doesn't. In concert with other elements, astringency can act as a wonderful counter to big, blowsy flavours, coming through at the finish to tie everything up neatly. On its own, in the wrong place, it can be horrible. The best description of astringency I ever heard was from my American friend Jenny, a supertaster, who described the experience of biting into an extremely tannic perry pear as 'like being punched in the face by a carpet in a really awesome way'.

The prickle of carbonation is a contentious one in some corners of the beer world. Sometimes considered a taste, the consensus now is that, given that carbon dioxide is tasteless, 'fizziness' is a trigeminal sensation. It's one that can still transform your experience of a beer or soft drink. The tingle it creates on the palate is a big part of the perceived refreshment value of lager – I often describe it as a 'bite'. Drink a fizzy drink too quickly, and it can start to become painful. But without any carbonation at all, any beer tastes quite different.[13]

Take another sip. Think about how the beer feels, rather than just how it tastes. We generally refer to the sum total of trigeminal effects as 'mouthfeel'. That's a term I only used to hear in advertising agencies, but it's now entering common parlance. It's an important part of the beer-drinking experience: it includes the thick, sometimes chalky body of juicy IPAs and pale ales. It plays a part in our perceptions of 'drinkability'. Now that low- and no-alcohol beers no longer taste like a rat's arse, mouthfeel is increasingly the only part of their flavour profile that gives them away. Alcohol content has what we tend to call a 'weight' on the palate, which we only notice when we drink something that tastes just like a beer, except that weight isn't there.

Temperature

There are all sorts of other factors that influence our perception of flavour, and we'll be exploring some of those soon. But while we're sticking with the physical dynamics of flavour perception, the one we need to mention here is temperature.

This is a long-running debate in the beer world. While what we perceive of as average room temperature is steadily increasing, the temperature at which we think drinks should be served is decreasing. Different drinks come with different serving temperature recommendations. Red wine should be warmer than white wine, and ale should be warmer than lager. But 'warm' is a tricky word in this context. There's still a persistent belief – among both its detractors and some of its fans – that cask ale should be served at 'room temperature'. This dates back to when our rooms were up to ten degrees cooler than they are now. When I was a kid, it was getting warmer than it had been, but the average room temperature was still only 18°C. It's now about 24°C. Cask ale should be served at cellar temperature, which is 11 to 13°C, and it's served at this temperature by anyone who knows how to look after it properly.

Why do we have these temperature bands for different drinks? Because colder temperatures mask some flavour attributes, and allow others to come to the fore. When coffee is served cold, it seems to taste more bitter. The wine thing comes partly from the fact that chilling wine enhances our perception of acidity, which you want from a white wine, but not necessarily a red. Meanwhile, sweetness seems to decline at cooler temperatures.

Joking aside, if you take a beer where the brewer's 'serving suggestion' is that you drink it ice-cold straight from the bottle, perhaps with some fruit in the top, try drinking it instead from a glass at the traditionally recommended serving temperature for lager (between 4 and 7°C), you'll instantly realise why the marketers do everything they can to prevent you from tasting it properly.

Why do we like some flavours and not others?

We enjoy the flavours of food and drink because it's essential to the survival of the human race that we do. From an evolutionary perspective, anything that's a common characteristic across a species is there because it helps that species survive.

The forerunners of Homo Sapiens – our common ancestors with apes – lived in forests and ate a fairly monotonous diet.

'What are we having today?'

'Fruit and leaves.'

'Ooh, nice. What about later?'

'More fruit and leaves.'

'Excellent. Say, do you know what I fancy at the weekend, as a treat?'

'More fruit and leaves?'

'More fruit and leaves.'

They all ate the same, and that was pretty much all they ate. If it wasn't safe or nutritious, they'd have gone extinct. But because it *was* safe and nutritious, there was no need for a complex appreciation of different flavours. So long as they were basically OK with fruit and leaves, they were good. Today, animals that still have these monotonous diets lack some of the taste receptors we humans have.

But at some point, our forebears came down from the trees and began to explore the savannah. There weren't as many leaves, but for omnivores like us, there was a huge, diverse, new range of potential foodstuffs. These were of variable nutritional value. Some were poisonous. So, we evolved a more sophisticated appreciation of taste and flavour, which rewarded us for eating what was good for us by giving us pleasure, warning us against eating what might be bad for us, and giving us the equipment to tell the difference.

Sweetness is the taste of sugar and, to some extent, fat. This represents a high concentration of energy, so we're predisposed to like it. Until very recently, humans didn't have access to readily available stores of energy all year round. So, when we taste sweet stuff, the brain releases endorphins telling us to eat more of it while we can, so we can store it as fat to use later. This is why some people gain a 'second stomach' when dessert is served after a heavy meal. Your brain is telling you that you are full, can't eat another thing, so you don't order a dessert. Then, when your partner insists you have just a taste of their sweet, glistening pudding, suddenly you have room and eat half of it and they get upset. I don't know if it's the same process when it happens the other way around with chips.

We'd also die without salt. And like sugar, we haven't always been able to get a lot of it. So, salt enhances our appreciation of other foodstuffs, increases our appetites, and makes us want to drink more, hence the universal presence of salty snacks in pubs and bars, whatever guise they may take.

Many compounds that are toxic to us present as being bitter, so we're programmed in evolutionary terms not to like it as a taste. One of the changes that happens to pregnant women in relation to food and drink is that they become more sensitive to bitterness. One hypothesis is that this is a protective response at a time when the baby's major organs are first forming, and the mother is highly sensitive to even low levels of toxins.[14] This could also be why children don't like eating brassicas and why beer tastes horrible to most people the first time they drink it. As Winston Churchill supposedly said, 'Most people hate the taste of beer – to begin with. It is, however, a prejudice that many people have been able to overcome.' The fact that many of us eventually do learn to love the bitterness of beer may be due to the confusing presence of bitterness in some things that are good for us. More likely, to me, is that when we start to associate a particular taste or flavour with pleasurable times, our brains tell us that we do like it after all.

It certainly works the other way round, with flavours we begin to associate with unpleasant experiences.

Sourness is an intriguing one. It's the taste of fruit that's gone rotten, so we're less well-disposed to it. But we don't dislike sourness as much as we do bitterness. This just might be linked to a relatively new theory which suggests there's an evolutionary advantage to enjoying a few drinks.

The 'Drunken Monkey' theory, posited by Dr Robert Dudley of the University of California at Berkeley in 2004, is based on the fact that when fruit is ripe, yeasts start to attack it and ferment the sugars into alcohol. The smell of ethanol carrying in the air acts as a kind of signal flare to primates that here is an abundant source of sugar. Animals that like the taste of booze – and humans are not the only ones – get wind of the smell and get to the ripe fruit, before animals that cannot process or don't like alcohol. Dudley presented this theory as a way of explaining the prevalence of alcohol addiction among humans. But surely that inverts the main hypothesis, which is that creatures that enjoy booze are more likely to survive than those that don't.

Overall, our appreciation of flavour helps us survive. It makes eating and drinking pleasurable. It helps us differentiate between foods that are good and bad for us – or at least it did until the modern age of abundance and ultra-processed foods. It also helps us digest food better. When your mouth waters in response to appetising smells, this is just one example of how our appreciation of flavour triggers positive actions in our digestive system.

This all explains why we enjoy flavour as a species overall. But why do some of us love some flavours and hate others, while for other people it's the opposite way around?

Evolution moves a lot slower than societal or social trends. Our individual tastes in food and drink are influenced by cultural, personal, historical, contextual and marketing factors that do nothing to change the way we taste – they can't change the complex interplay of aromas, taste and mouthfeel – and yet they can completely change our preferences or even our perceptions of what something tastes like. We'll talk more about this later.

The complexity of flavour

So, our perception of flavour is the combination of taste, two types of aroma, and a bunch of different sensations we can usefully call mouthfeel. The fact that flavour is this complex helps explain why it's so difficult to describe. Sight is sight, and hearing is hearing (or so we think) – flavour is far trickier to pin down.

On top of that, much of what we've explored here is still not unanimously signed off by the food scientists, biologists, psychologists, neuroscientists and philosophers who are all currently researching this field. Some are now arguing that tastes can be smelled, or that you can't have a taste without an aroma, or vice versa. There's an emerging argument that all tastes are actually flavours, that they're essentially the same thing, which would mean that after all this palaver, the average person was right all along. The fact that the brightest minds studying flavour still haven't quite agreed on what it actually is should be of some comfort to the rest of us.

Why is all this relevant to beer and music pairing? Because it starts to lead us to a scientific explanation of how it's possible for the sound of a song in your ears to alter your perception of the flavour of the beer in your mouth.

Tasters like to say that all five senses should be involved in a good tasting. We've already established that our perception of flavour itself is a combination of at least three of them, and that the brain can be tricked into mistaking aroma for flavour, mouthfeel for taste, and so on.

The key to all this is that information about aroma goes into the deep, limbic brain very quickly, at the same time as taste signals from the oral cavity are transmitted to the brain along cranial nerves. We form our overall impression of flavour by bringing these signals together. In other words, our perception of flavour is formed not in the mouth or nasal cavity, but in the brain itself.

Two hundred years ago, the French gastronome Jean Anthelme Brillat-Savarin – arguably the man who invented modern food and drink writing – finished his own description of the tasting process by saying, 'Finally, the reflective sensation is the opinion which one's spirit forms from the impressions which have been transmitted to it by the mouth.'[15]

We can see he wasn't quite right, but he was ahead of the game. Flavour notes are just chemical compounds until our brain takes all the information and puts it together to form an overall impression. The brain frames this impression within a context of established preferences, tastes, beliefs and previous experiences. And unless we're doing this in a controlled scientific experiment, it tries to do this at the same time as it is also trying to make sense of what we can see – and what we can hear.

Flavour and beer

As we'll see later, the vast majority of the work that's been done to understand the relationship between flavour and sound has used wine (and classical music). Only the various researchers can tell us exactly why this is, but I have my suspicions. Even now, seventy years after the birth of rock and roll, when almost everyone still alive has been part of a rebellious generation of teenagers, pop and rock music are still not taken as seriously as classical music within the space of academic study or true discernment. And I know from personal experience that beer is not taken as seriously as wine by anyone other than beer geeks like me.

I actually drink wine as often as I drink beer. I have absolutely nothing against it. It's my off-duty drink. I deliberately don't analyse it or learn more about it, because I want one drink that I can enjoy like everyone else does. I don't think beer is better or worse than wine – both can be great or awful. I've never argued that 'beer is the new wine' or anything like that. The older I get, the more often I find myself saying, 'You are allowed to like more than one thing.'

I love using beer as the medium to talk about flavour, not because I think it's better than wine *per se*, but because it's better than wine at this.

Wine is great if we want to explore sweetness, acidity and tannin. Beer can do all of these quite easily, so long as you can get your hands on a broad enough selection of styles (milk stouts, Flemish red ales and wooden-barrel-aged beers respectively.) But it can also do bitterness (which, as we now know, is a form of dryness, but not the same as tannic dryness.) There's plenty of umami, especially in aged beers. You can even find the odd salty beer.

That completes the basic tastes. Smell-wise, we've already covered the unparalleled array of aromas that hops bring. But don't let that get in the way of malt. If the dominant aromas in hops cover grass, hay, white pepper, pine, citrus, and tropical fruit, malt adds the possibility of nuts, biscuit, toffee, dried fruit, forest fruit, caramel, coffee, chocolate and tobacco. Then there are the aromas yeast creates during fermentation, which could, depending on the yeast and the beer style, bring to the party some fresh bread, spice, butterscotch, more fruit, banoffee pie, even a farmyard. Think about the combinations all these different notes make possible. Then, add water chemistry, with different salts in the water bringing out or pushing down certain flavours. And there's also barrel ageing, with the potential to add tannin, oak and vanilla, and simple ageing itself, where some flavour compounds break down and new ones emerge: leather, port, madeira, soy sauce are just some of the flavours that time can bring.

Also add to that the trigeminal effects of carbonation, how that varies from cask ale to commercial lager, the varying weight of alcohol, the thickness of some hazy IPAs. If you really love the taste of wine, there are beers that can taste just like that, too.

This is why I never get bored of beer. And it's why beer is the best possible drink there is to explore the full vista of how flavour works. If you're then going to experiment with how flavour interacts with other senses, why would you choose anything else?

How hearing music works – and why it's a lot more complicated and uncertain than you probably thought

I've been studying flavour for many years. It fascinates me as a writer, as well as someone who evaluates and enjoys drinking beer. However, as a music lover, I've always been happy just to possess a good sense of hearing (and excellent taste in music, obviously) without needing to understand the science behind it. We don't need to know all about the workings of an internal combustion engine to enjoy driving, or the mechanics of photosynthesis to enjoy eating salad. Until now, I've kind of bluffed the science of hearing through my events, just showing a diagram of the ear and going, 'What about that then, eh? Crazy!'

I knew I needed to tackle it properly for this book, and I was fairly blasé about it. Tasting is a complex interplay between three different senses, and our brains don't really understand that this is the case. Hearing is straightforward. It's one sense. There's a noise. We hear it. With our ears. Simple.

Yeah, so that was totally wrong.

Hearing is one of the 'noble' senses, with far more attention devoted to it over the centuries than the perception of flavour. But there are still quite a few fundamental aspects of how we understand what we can hear that we really don't have much of a clue about. Also, I was horrified to discover that what we do know involves a lot of maths, especially algebra. I'll be avoiding that here wherever I can.

Like flavour, from a scientific point of view, understanding hearing requires a multi-disciplinary approach. We can break it down into:

- What is sound and how is it made?
- How do the ears sense and process sound?
- How does the brain interpret the information the ear sends it?
- Music is far more than just sound or noise. How does that work?

This means it's a subject that interests physicists, biologists, psychologists, philosophers, neuroscientists, and musicologists, who each have to understand what all the others are saying. Which, of course, means that the finer points are still being argued over. Here's what we know for sure. Well, sure-ish. It covers the basics, because once you understand those, it's much easier to understand the work that's being done to understand how hearing and flavour perception work together.

What is sound and how is it made?

Sound needs a medium to travel through. As we all know, in space no one can hear you scream. That's because sound is all about moving molecules. If you're in a vacuum and there are no molecules to move, there can't be any sound. Clearly, this is a blow to all fans of the sound engineering in *Star Wars* dogfights between X-Wing and TIE fighters, not to mention the photon torpedoes on the *USS Enterprise*. But if you're that into space and science fiction, you probably know this already, and take great enjoyment in correcting noobs about it at ComiCon.

For most of us, most of the time, the medium through which we experience sound is air. It also travels through liquids and solids, but if you're not careful this can all get extremely complicated very quickly, so let's just focus on air for now.

The molecules in air travel randomly at very high speeds, constantly colliding with each other. Sound is caused when something vibrates, and thereby disturbs the random meandering of the air molecules close to it. This vibration could be caused by a knock on a wooden door, a stick hitting a drum, a piano hammer hitting a wire, or two pint glasses striking each other. The vibration of the material means it moves outward, and then inward as it tries to recover its original shape. It goes back and forth in slightly smaller increments, until it returns to its stationary state. When it moves outwards, it pushes molecules in the air and compresses them into bands of high pressure. When it moves back inwards, it creates areas of low pressure, called rarefaction. As the object moves backwards and forwards, it creates a sequence of compression and rarefaction, bands of high-pressure and low-pressure air molecules. The molecules that have been compressed and rarefied hit other molecules next to them, and they hit the molecules next to them, and so the wave travels through the air (or any other medium). If you plotted the changes in air pressure on a graph, it would create a sine wave. These waves are how we visually represent sound and explore its different attributes.

The harder something is hit, the more it will move outwards and back inwards, meaning the air is compressed and rarefied to a greater degree, meaning the peaks and troughs of the wave will be higher and lower respectively. We'd perceive the resulting sound as louder. But that's too easy to understand, so scientists call it *amplitude* instead. If it's louder – if the waves' peaks are higher – the sound has higher amplitude.

Different materials vibrate slower or faster. Imagine a guitar string in extreme slow motion. When it's plucked, it goes in one direction, then back past its starting position and out in the other direction, then returns to its starting position once more. This is one cycle of movement. If something takes one second to do this, we measure it as vibrating at 1 Hertz (Hz) after Heinrich Herz, the German theoretical physicist who first transmitted radio waves. If something vibrates slowly, you might hear it as a series of clicks, like a metronome. Above a few Hz, the vibrations are so quick we perceive a continuous noise, or *tone*. The faster the vibration, the higher the tone. Because it's all about the speed at which the thing is completing each cycle – the rate at which its vibrating – we refer to this as the *frequency* of the tone. I used to think frequency was a complicated concept, but it's the same use of the word as any other.

How many times does it go through a cycle in one second? The typical human ear can hear frequencies between a range of 20 and 20,000 Hz – the sound of something vibrating between 20 and 20,000 times per second – but we don't necessarily register frequencies in the higher range as sound. We're most sensitive to frequencies between 2,000 and 5,000 Hz.

We tend to refer to the sound of this frequency as the *pitch*. Technically, pitch is not a property of sound waves, but a mental construct we use to make sense of them and put them in order, and we usually apply it specifically to music rather than just sound. High frequencies produce high-pitched sounds, and vice versa.

There's a lot more to sound than amplitude and frequency – or pitch and volume, if we want to keep things simple. Like light waves or other waves, sound waves are reflected, refracted, and diffracted, and exhibit interference. We could talk about velocity and transverse versus longitudinal waves. But let's not. All we need to understand for now is that sound consists of waves of higher and lower air pressure, created by something such as surface being hit, which moves columns of air, at higher or lower frequency and volume.

This provides us with a straightforward, factual answer to a question that's often presented as if it's some great philosophical conundrum. If a tree falls in a forest and there's no one there to hear it, does it make a sound?

Sorry, philosophers, but if only you'd listened: the simple answer is no. Those waves of compressed and rarefied air are only perceived as sound once they hit an ear or some recording equipment. What we think of as sound only exists inside our heads. If there's no one there to hear it, it's just air molecules marching across space.

If there are ears there to hear it, well, that's just the start of the fun.

How do the ears sense and process sound?

The earlobe is a perfectly easy thing to understand. We can see it and recognise it, and we refer to it simply as 'the ear'. But now we're in hearing science world, it's not called an earlobe anymore; it's a pinna. This is the Latin word for 'fin' or 'wing', so unless you're still a child looking for a cruel schoolyard nickname, I really don't see what use that is to anyone. Which is probably why it's also known as the auricle or auricula. Technically the lobe is just one part of the auricle.

Auditory system

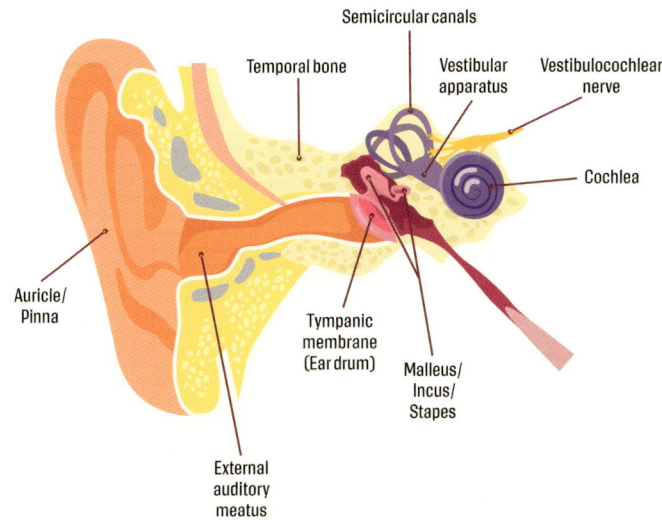

Sometimes academic jargon reaps unexpected rewards, however. The fin or wing directs sound waves into a narrow passageway called the ear canal. The ear canal is also known, magnificently, as the 'external auditory meatus'.

If you focus on just the mighty head fin and the most excellent EXTERNAL AUDITORY MEATUS, you can see that together they form a funnel, collecting sound waves, amplifying them, and directing them internally towards the ear drum. Given that it behaves like a drum, here, finally, is a name that is more appropriate and understandable. (Apart from when some of the buggers insist on calling it the 'tympanic membrane' instead.)

In response to the sound waves hitting it, the eardrum vibrates, setting off more vibrations in three tiny bones in the middle ear. These bones are called the malleus, incus and stapes. Your school biology teacher probably tried to get you to remember this by calling them something completely different – the hammer, anvil and horseshoe. To be fair, the proper names are the Latin words for these English nicknames.

Then this snail comes along. *It does!* Look at the diagram. Except it's not really a snail; it's the cochlea. Why is it called the cochlea? Because that's the Latin term for snail shell. Even when you try to take the piss, these people are way ahead of you.

The cochlea is filled with fluid, through which the vibrations continue. All along the cochlea is a membrane covered in tiny hairs. And although I now sound like Danny the Dealer from *Withnail and I*, these hairs turn vibration into electronic signals, which are carried along the cochlear nerve to the brain.

Guess what? Scientists still don't understand how this actually works.

There's also a round window and an oval window in there somewhere, as well as the vestibular system, which is not a distant galaxy where the rebel alliance is hiding, but a set of organs within the ear that are responsible for your sense of balance.

But let's come back to those hairs that are your aerials, man. They're not just arranged at random. They're in a pattern. Different hairs are 'tuned' to detect different pitch, and they're arranged in order from high pitch – 20,000 Hz at the outside opening to the 'shell' to low pitch – 20 Hz at the middle of the spiral. As we get older, the hairs at both ends grow stiff and can't vibrate any more. That's why people over 40 lose some of the range of pitch we can hear, and why pubs and bars that are full of hard surfaces and tinny music, with nothing to absorb the sound, become really annoying.

How does the brain interpret the information the ear sends it?

Like flavour, our perception of sound is formed in the brain. In our heads, those waves of air pressure are interpreted not just as having properties of pitch and volume; we get a sense of direction and spatial awareness; where it's coming from and how close it is. We can screen a cacophony of sound and pick out the relevant bits, separating the sound of a band from the sound of the audience cheering and any background noise, for example.

As I'm writing this, I just went to the kitchen to make a cup of tea. The washing machine was on. Since we had it repaired recently, it shakes quite violently during the spin cycle. I could hear this from the hallway outside the kitchen. From the sound waves hitting my ear, I could tell that the shaking noise was indeed coming from the kitchen. I could identify that the source of the noise was the machine shaking across the floor, rather than just being completely stationary as it should be, and that it was shaking quite violently. I could also hear a water bottle rattling on the top, which I'd forgotten to move when I put the laundry on. My memory filled in that it was a water bottle, but I could recognise it from the specific clattering sound, which was of a higher pitch than the shaking of the machine. I then heard a crash, which I recognised as the water bottle falling to the floor. I knew it was the floor, because I could tell it was the sound of a hard object hitting a hard surface. The reverb from the crash told me that the water bottle was empty rather than full. The timbre of it told me that the water bottle had not broken, and the staccato rhythm of the sound told me that it bounced a few times before coming to rest, each bounce shorter than the last.

How do we decode a waveform hitting the eardrum into all this information, separating one detail from another? Academics say that this is a 'problem that is mathematically ill-posed', which is science for 'We don't know', despite decades of psychological, physiological and computational research.[16] The way the ear sends information to the brain doesn't work in the same way as microphones, amplifiers and earphones do, which has a straightforward mathematical explanation – if you can do maths.

What we do know is that the auditory nerve carries signals from the magic hairs in the snail shell to the auditory cortex, in the heart of the brain, where a 'tonotonic map' matches perfectly with the arrangement of the hairs. So, if a certain pitch enters the cochlea, the hairs that correspond with that frequency send signals to the tonotonic map, and the same pitch 'lights up' in the brain. Imagine a piano keyboard where each key you press lights up a different light in a light show. That's what happens in your brain every time you hear a sound. That's how important the pitch of sound is in human evolution.

While that's happening, another part does all the working out of where it came from, how loud it is, and so on. How? Let's ask Sir James Beament, biologist and author of *How we Hear Music: The Relationship Between Music and the Hearing Mechanism*. He writes, 'The [auditory] cortex is so complex that the most we may ever hope for is to understand it in principle, since the evidence we already have suggests that no two cortices work in precisely the same way.'[17]

Thanks, Sir James.

So, we might as well leave it there for now in terms of how the brain makes sense of sound. All we know is that it does, and that it's so sophisticated, it's beyond the comprehension of modern science. In an age where we think we know everything, it's humbling, and rather exciting to come up against the limits of human knowledge. Apart from anything else, it reinforces the importance of this section, Side One of this book. If we don't fully understand how hearing works, how can we possibly say with any certainty that it *doesn't* impact our perception of other senses, including flavour? Is that any less likely than the ear being responsible for your sense of balance? Or tiny hairs inside a sea shell converting vibrations into electronic impulses?

Music is far more than just sound or noise. How does that work?

The most definitive research that does exist on understanding the brain and sound has focused specifically on music, so this would be a good time to jump into the mysteries of great tunes.

There is no known human culture now or at any time in recorded history that hasn't had music. Whistles and flutes made from mammoth ivory and bird bones, discovered in caves in southern Germany, have been carbon-dated and shown to be 43,000 years old, placing them among the earliest ever human-made artefacts.

The materials that drums were from were more prone to decay and haven't lasted, but percussion instruments are thought to go back at least as far. The stereotypical caveman of cartoons and comedy sketches wasn't dragging mates back to the cave with a big club: they were seducing them with song.

We might assume the music of 43,000 years ago was pretty rudimentary, but we don't know that. The human brain was just as big as it is now, and possessed all the mental frameworks we now use for making music, even if we lacked sophisticated knowledge back then. But this leads to the question: what is music? Or rather, what's the difference between music and noise?

And you thought the definition of craft beer was problematic...

Unsurprisingly, there's not really a definition of music that everyone is happy with. French composer Edgard Varèse defined music as 'organized sound', particularly with reference to his own music, which explores timbre and rhythm and doesn't seem quite so concerned about melody. He spoke of 'sound as living matter', and of 'musical space as open rather than bounded'. He speaks my mind when he argues that 'to stubbornly conditioned ears, anything new in music has always been called noise'. But if you search for his own music now, and listen to it for the first time, you might very well think, 'Yeah, well. He would say that, wouldn't he?'

'Organised sound' has, nevertheless, found itself as the jumping off point for many definitions of music. It's good enough on its own for me. But not for others. While any kind of sound can potentially be included in music so long as it 'fits' within the overall organising scheme of the piece, some argue that the definition is too broad: isn't speech also organised noise? Isn't a fire alarm or police siren?

Inevitably then, people try to put extra caveats and conditions on it. *The Concise Oxford Dictionary* defines music as 'the art of combining vocal or instrumental sounds (or both) to produce beauty of form, harmony, and expression of emotion'. This is ridiculous. Not only would it require everyone in the world to agree on the same idea of what constitutes 'beauty of form, harmony, and expression of emotion', it also rules out any music using techniques such as feedback and distortion, and styles such as *musique concrète*, which uses recorded sounds, often on tape, often as a collage of 'found' sounds.

Having survived writing a whole book about the definition of 'craft beer', I'm not going any further with this one. Let's stick with 'organised sound' for now. After all, there are loads of songs with talking in them, and quite a few that use sirens and alarms. I wish you the very best of luck if you want to argue with Liz that 'Blockbuster' by The Sweet is not music.

The idea of 'organised sound' tees us up quite nicely for a quick romp through the major properties of music. If music is organised noise, what are the organising principles? Some of these are properties of the sound itself; others are internal mental constructs that we create in our brains to interpret and codify sounds. It's worth spelling out how the basic parts work, because the experiments on matching flavour and music that we're going to look at later often drill into specific aspects of music.

Pitch

In one sense, music consists of vibrating columns of air. The vibration might be caused by someone blowing into the tubes of a wind instrument; a stick hitting a drum; a piano hammer hitting a wire; or a guitar string being plucked. As we know, the vibrations create sound waves at different frequencies, and in music, we refer to these frequencies as *pitch*. We usually call what we hear a *note*. Strictly speaking – and it seems obvious when you think about the word – the note isn't the sound you hear; it's a notation on a piece of paper representing that sound. What we actually hear is correctly termed a *tone,* which is of course an anagram of *note*. Which is nice.

While frequency is a property of sound, pitch is part of the mental framework we use to conceptualise sound. The pitch of a tone only really becomes music when it's joined together with other tones to form a *melody*. We classify groups of tone together into a *scale* (from the Latin *scala*, which means 'ladder'). In terms of pitch, a scale spans an octave, so called because it has eight major notes: A-B-C-D-E-F-G. Imagine a piano keyboard: when we get up to G, the next note is A again, an octave higher than the first A. From there, we repeat the scale in higher tones. The second A may be a different tone, higher than the first. And the third and fourth As are of course higher still. But people with perfect pitch still recognise them all as As. There's a mathematical relationship between them: the second A is exactly twice the pitch of the first one, the third the double of the second, and so on. The lowest key on the piano keyboard, on its far left, is an A. The keys towards the left cause hammers to strike longer, thicker strings that vibrate more slowly. The frequency of the lowest A is 27.5 Hz. The strings get shorter and thinner as you go up the keyboard, causing vibrations at higher pitch, The eighth A on the keyboard, the third-highest key overall, vibrates at 3520 Hz – which is 27.5^8, or 27.5 doubled seven times.

Nursery rhymes – 'Twinkle, Twinkle Little Star', 'Baa Baa Black Sheep' – and Christmas songs such as 'Jingle Bells', only use one octave. They can be played on toy keyboards. They're simple to remember, sing and play.

A full piano keyboard has seven-and-a-bit octaves, which is the most of any musical instrument, which is in turn why it's so popular for composing music. Even though A is the lowest key on the keyboard, C is the note that usually starts an octave. If you go from any letter up to the same letter again, it's still an octave. If you start with C rather than A, that means you're playing in the key of C.

In each octave, the white keys are the *major* notes, A to G. But the black keys throw in a handful of sharp and flat variations of those notes – tones which are higher than one major note, but not as high as the next one. So a full piano octave actually consists of twelve keys, even though it's named after the eight major notes, just to make sure you're still paying attention. The black keys, falling between the major notes, are *minor* notes. In Western music, we associate major notes with happy, confident tunes, and minor notes with sadness or melancholy.

This scale of twelve notes from A to G is known as the chromatic scale. It's the basis of Western music. It sounds 'right' to us because most of what we've heard since we were born (or even before – see below) has been composed using it. But a scale is an intellectual and cultural construct. Other scales do exist. They always span an octave, but may deploy a different number of notes within it, such as five (the pentatonic scale, often used in folk music and especially Asian music) or seven (a heptatonic scale, used in music such as Indonesian gamelan).

Pitch is therefore the first key building block of music. When the first experiments exploring the relationship between music and flavour were carried out, they focused exclusively on pitch.

Timbre

Timbre (rhymes with *amber*) is the voice of an instrument – or a voice, I suppose. If you and I were speaking in the same monotone, at the same pitch, timbre is the quality that would make your voice sound different from mine. Similarly, if a violin, a trumpet and an electric guitar all played the same note in tune with each other, it's the quality that would allow your brain to identify which instruments were playing, and pick out one from the others.

When you pluck a guitar string or blow into a saxophone, it doesn't vibrate at just one frequency. Remember the multiples of A at different frequencies? If you were to hear, say, a saxophone note, being played at 220 Hz, you wouldn't just hear it at that frequency. You'd also hear multiples of it too, at 440 Hz, 660 Hz, 880 Hz and so on. These multiples are called *overtones*, as in, over and above the main tone. Depending on what the instrument is made of, how big it is and what shape it is, different overtones will be more or less prominent in one instrument compared to the next. The overtones from the sax will sound different from those of a guitar, even if they're both playing the same 'fundamental tone' of 220 Hz. Because of this relationship between tones and overtones, the timbre of a single instrument or voice will change if it's making high notes compared to low notes.

Listen to 'Rooting for You' by London Grammar, essentially a straight-line run-through of singer Hannah Reid's extraordinary vocal range. She is a contralto, the lowest female voice, and in the song she climbs from her deepest note to a high falsetto and back again. The range is astonishing in itself, but the timbre – the actual *character* of her voice – changes markedly as she goes up and down.

Daniel Levitin is a musician, songwriter, sound engineer and producer who went back to school and became a neuroscientist and cognitive psychologist. He's also a wonderful writer. Understandably, I'm a little bit in love with him. He calls timbre 'the most important and ecologically relevant feature of auditory events,' allowing us to distinguish between the roar of a lion or the purr of a cat; or the voice of a friend from someone you're trying to avoid.

Rhythm and pulse

Rhythm could possibly be the oldest element of music. When we think of rhythm in popular music, we think of the *rhythm section*, of drum and bass pounding, keeping the other musicians in check. But rhythm isn't just a part of the band, and it's not just the beat: it's a property of the music itself, namely a pattern of notes of different lengths. The length of a sound is called the *duration*. A pattern of sounds of different durations makes a *rhythm*. The drum machine intro to New Order's 'Blue Monday' is an excellent example of this, as well as being excellent in every other imaginable respect. This is different from the *pulse* of a song, which we also called the *beat*. In a pulse, sounds are the same length. In a rhythm, the sounds are different lengths.

Rhythm is what makes the hook for the Beach Boys' 'Barbara Ann' – the first seven notes all at the same pitch, with only the rhythm varying. Rhythm gets a grand day out on 'The Sunshine Underground', a track on the album *Surrender* by the Chemical Brothers. The role of rhythm and melody are reversed: the melody stays constant, essentially keeping a rhythm, while rhythm and pulse go all over the place.

Melody

Melody is the main theme of a piece, an agreeable combination of notes. For me, it's one of the most powerful and fascinating examples of the cultural importance of music.

Our pleasure from melody follows a similar line to our love of comedy: the set-up builds tension, and then the pay-off releases that tension. In comedy, we laugh because our expectation has either been met or subverted, or a combination of both.

In music, the subversive part is a lot harder to do successfully. Let's go back to those simple, one-octave nursery rhymes again. 'Twinkle Twinkle Little Star' is sung to an 18th-century French melody, 'Ah! vous dirai-je, maman' ('Oh! Shall I tell you, Mama'). The melody is also used in other nursery rhymes, including the 'ABC Song' and 'Baa, Baa, Black Sheep'. It rises from the first part. The middles part kinds of leaves you hanging. And the final part takes you back to where you started. The middle part creates tension – it sounds like it's not finished yet – and the final part reassures us that everything is OK. Imagine if the melody finished on the bit where you sing 'Up above the world so high/Like a diamond in the sky'. Even from a very young age, you know that's not 'right'. The tension hangs. You know it's not finished. We work on patterns. The tension and release in this simplest of melodies tees up the use of key changes, shifts from major to minor, dissonant, unexpected notes to suggest something is wrong, and so much else that allows music to manipulate our emotions by messing with our sense of what it 'should' do next.

...And the rest

There's a lot more to music that we don't need to go into in as much detail. The mechanics of pitch, timbre and rhythm give us all we need to understand the science behind pairing music with flavour. But there are some other terms that will occasionally be mentioned. *Tempo* is the overall speed or pace of a piece of music. *Reverberation* gives us clues as to how far away the source of the music is, based on the echoes in space. *Harmony* is the relationship between the pitches of different tones being played or sung simultaneously. And then there's *space*: the bits where none of this is happening. Miles Davis used to say that the most important part of his solos was the 'air' he left between notes, allowing the listener to build up expectation.

The remarkable thing about our relationship with music is that we are able to separate and identify all these attributes and more, individually. One can be varied without the others, and we can pick it up.

Music and the brain

Our ability to identify different instruments and different components of music is, in the words of the author of a frankly impenetrable text book on the subject, 'far beyond any physiological explanation at this time'. I wonder if this is why scientists seem to deliberately make it so difficult to understand what we *do* know – perhaps it's a form of over-compensation.

We do know quite a bit about what music means to us, and can speculate on why, even if we don't know how.

We start listening to music in the womb. The auditory system of a foetus is fully functional by the time it is twenty weeks old. Experiments have proved that babies hear songs their mothers played while pregnant, and they recognise and prefer these songs to new ones up to a year after being born. Music helps babies learn, and mothers know this instinctively. Many mothers sing to babies to help them sleep, or even make them smile. Even if we can't sing, we at least exaggerate the intonation of our voices into a kind of sing-song, using a slower tempo, higher pitch, and a more exaggerated range than if we're talking to adults or older children. We do this without any instruction to do so.

It turns out that exposure to music at an early age is closely linked to the development of speech. This is when our brains are like sponges, primed to absorb as much as we can from the environment around us. If a child is not exposed properly to music or language before a certain age (sometime between eight and twelve) they will never acquire normal music or language skills. This is because the synapses in the brain are still developing. Try learning a music instrument or new language after the age of twenty, when this development is complete, and you'll never be as good, as natural at it, as someone who learned as a child.

I believe we're born with a longing for music. We'll be talking later about context, which is a huge part of our relationship with music. But even before we've had a chance to form any context, music still moves us. One of my earliest memories, when I was two or three, is the music on the opening credits of the TV show *The High Chapparal*, which ran between 1967 and 1971. As soon as I heard it, I burst into tears. I loved it, but it moved me in ways I simply couldn't understand. Mum and Dad said I was overtired. I got put to bed, and was always put safely there before the programme started after that. Listening to it as I write this, the key change still chokes me. It paints emotions with a very broad and obvious brush. Which is why, as a toddler, those emotions hit me in the gut, hard.

Typically, the music that means the most to us throughout our lifetimes is the stuff we get into in our teens. Around ten, even children who weren't that into music before start to take an interest. In the emotional storms of adolescence, the music we love means everything, because it soundtracks and seems to speak to emotions that feel so huge, that we're struggling to deal with.

Many people simply stop exploring new music after early adulthood, when our brains have finished developing fully. Nothing can match the music of a few years ago, which still pulls at your heartstrings. New stuff just doesn't seem to mean as much, somehow. By the time you hit middle-age, the adolescent girls sobbing at the live appearance of the latest boy band seem like alien creatures. Because you never fantasised about marrying Kim Wilde or meeting Liam Gallagher in the pub and him suggesting you become best mates. Hell, no.

Music also lodges in our memory extraordinarily well. Most of us can remember large numbers of songs, and we can remember way more tunes than we can lyrics. Music often outlasts most other memories in people suffering from Alzheimer's. It uses many other parts of the brain as well as those that look after speech. People can remember, sing and even play the music they loved at the age of fourteen long after they've lost the ability to speak or recognition of the faces of loved ones.

Music and meaning

That you even picked up this book, let alone read so far, means I probably don't have to convince you about the enormous emotional power of music. Music makes us happy. It makes us cry. We use playlists and albums to both enhance and change our emotional state.

For some of us, there's joy even in sad music. I find that if I'm angry, stressed, or upset, if I play a song that embodies that emotion, it evens me out, because the external environment reaches equilibrium with my inner turmoil. The furious squall of noise of My Bloody Valentine's 'Only Shallow' can calm me down, and the tear-jerking sadness of Billie Eilish's 'What Was I Made For?' can cheer me up.

For most of our history, music hasn't just been something to listen to. It's also something to be performed and shared. When doing heavy work, be that bringing in the harvest, hoisting the sails on a ship, or enduring slave labour on a cotton plantation, singing songs would help pass the time, and also establish a rhythm for coordinated actions. Folk traditions around the world – not least the Blues – arose out of this kind of communal singing. The separation of audience and performer, and the elevation of performers to star status, is a relatively recent innovation that predates recorded music, but was massively accelerated by it. In the nineteenth century, the 'free and easy' – a communal pub sing-along – was the forerunner of the 'turn' in a working men's club, and of music hall, where the first British 'popular music' was born from a mish-mash of international folk traditions. It never really disappeared: in many pubs and clubs across the UK, the karaoke night remains a guaranteed way to pack the venue. People who attend church still feel the benefits of communal singing. Obviously, the exercise helps improve your lung function. It makes you feel happier and more connected to others. It has also been found to help boost immunity, improve the speech of people who have suffered ailments such as a stroke or Parkinson's disease, relieve stress, improve posture and memory, and, basically, help you live longer.

We should all sing more. It's good for us. Who cares if you're tone deaf? If we can dance like no one is watching, we can sure as hell sing like no one is listening. I'm not sure playing an instrument is as much good for your physical health, but it's certainly good for your wellbeing. When I was being an awful lead singer in a university band, our guitarist was having a much better time of it. He'd sit propped up on his bed strumming chords, coming up with fragments and riffs, and I'd be fascinated at the trance-like state he fell into. His breathing when he did this was that of someone in a deep sleep. Too many people give up because they're 'not good enough to make it' as a musician. I envy people who can play an instrument, and do so just for themselves.

Even if we don't sing or play for ourselves, few things can match the euphoria of going to see those singers, bands and DJs who are good enough to make it. We're in communion with the rest of the audience and the band themselves. The audience feeds off the energy projected by the band, and the band in turn feeds off the audience. The feedback loop builds to levels that are far greater than the sum of the parts.

It helps that love is by far the most popular topic for songs. Love inspires us to write, perform and listen to music, and in turn music helps us understand and deal with love. Love is timeless and universal, like music itself. We've all loved someone – with the possible exception of people like Donald Trump and Elon Musk, and you can even include them if you want to count narcissistic sociopathy as a form of self-love. Love is a complex emotion that, for something so universal and powerful, is tricky to pin down. Songs ask 'Is this love?', and singers declare that they 'Wanna know what love is'. There's the euphoria of falling in love, and just as powerful is the heartbreak of losing the one you love. There's also more than just romantic love: love for parents, family, some imagined god or deity, even a mutant killer rat.[18]

Whole libraries have been written on the subject of music and emotion. For the purposes of this book, which has a lot more ground to cover, I'll leave it there for now, but hopefully illustrate the point emphatically in the choice of songs for Side Two. If you want to read one book on the topic that will inform, entertain and move you in equal measure, make it *The Sound of Being Human*, by Jude Rogers. Having listened to music all her life, Jude chooses particular songs that soundtrack key moments of love, loss, joy inspiration, and so much more, and translates her experience into the universal.

Of course, all this is open to interpretation. Music is gloriously subjective, and you might not like the same songs as me or Jude. As Iggy Pop said in a 1977 interview on CBC, sampled by Mogwai on their track 'Punk Rock': 'What sounds to you like a big load of trashy old noise is in fact the music of a genius: myself.'

To my parents, my music was a load of trashy old noise; and although I'm not a parent myself, if I did have kids, the contemporary music that I assume they would be listening to sounds like trashy old noise to me.

Sometimes, our reactions to music we don't like go far beyond dismissing it as rubbish. In some countries, even today, people are persecuted or even imprisoned for singing or even listening to 'forbidden' music.

The idea of 'rock and roll' as 'The Devil's music' has deep roots. The Catholic Church has always had a problematic relationship with a decent tune. In the fifth century, St Augustine wrote that singing beautiful melodies in praise of God would lead 'weaker minds' to be 'stimulated to devout thoughts by the delights of the ear,' which was a good thing. However, he confessed that sometimes he enjoyed the tunes of the Psalmody more than the lyrics, and he saw this as a grievous sin. In the Middle Ages, the church banned polyphony (music that contains two or more independent melody lines simultaneously) for fear that it would encourage people to doubt the unity of their god. (This, from the theologists who thought that the idea of god the father, god the son and god the holy spirit was a perfectly straightforward and clear way to explain a single deity.) Later, they banned the musical interval of an augmented fourth, the distance between C and F-sharp that's also known as a tritone. This interval, which Levitin helpfully explains to us novices as the music ground covered by Tony in *West Side Story* when he sings the name 'Maria', was considered so dissonant that it was perfectly obvious to any sane person that the Devil must have written it, and it was dubbed *Diabolus in musica*. It's still referred to as 'the Devil's chord', and is used to denote evil in everything from the musical *Phantom of the Opera* to a recent episode of *Doctor Who*.

Music, then, is as powerful as it is mysterious. It's also as tricky for us to fully understand as flavour is – maybe even more so. If music can do all this, the idea that it can make beer taste better almost seems trivial and obvious by comparison. And if we don't really know how music works in the brain, who can possibly say, with any degree of confidence or authority, that music definitely *doesn't* make beer taste better?

Now we've explored, to the best of our ability, how music and flavour work separately, the next part of understanding how they work together involves digging into how we experience both in our everyday lives, and how this adds further levels of meaning and complexity – and also starts to 'change' what the evidence of our senses perceives in isolation.

Confounding factors – how almost everything else affects your perception of what you think you're tasting and hearing

You know how you appreciate a cup of tea a little more if you're outdoors, how Guinness tastes better in Ireland, and how you loved Dorada, the local lager, so much on holiday in Tenerife that you brought a couple of bottles home, opened them under rainy skies and they tasted awful?

We often dismiss these observations as sentimentality or suggestibility, beating ourselves up for being gullible. But we're not being stupid: the right drink really does taste better in the right situation.

Context is everything. All that sensory information we discussed in the last couple of tracks isn't just splashing onto a blank mental canvas. It's hitting our brains at the same time as the information coming in from all our other senses. At the same time, our brains refer to memory, habit, previous experience and a whole host of other tricks and tools to help try and make sense of it quickly. We use what we already know and believe to form our opinions and reactions to what we experience in the moment.

This makes the objective analysis of flavour – and to some extent, music – very difficult indeed, supposing we even wanted to do that, as opposed to just enjoying more of what we know we already like.

When constructing experiments, scientists talk about 'confounding factors'. These are any variables that might interfere with the experiment; external conditions that we're not looking to include or measure, which can alter the results of what we are trying to measure. In my stage show, I start by saying that there are only two potential confounding factors that could interfere with what I'm trying to demonstrate. One is everything in the room

we're in, from the ambient temperature to the colour of the walls
to the seats you're sitting on. The other is every single thing that
has ever happened to you since before you were born, right up to
the point you walked into the room. To avoid confounding factors,
we should do this under laboratory conditions, playing pure tones.
But why would we do that when we can do it with great beers and
great tunes instead?

It's worth examining the confounding factors of culture, context
and individual experience more closely. While taste and music are
complicated enough on their own, they're influenced and percep-
tually altered by all this other stuff too.

Culture and ethnicity

Our ideas about flavour are shaped by the culture we live in. Cheese
is often perceived as disgusting by East Asians who are perfectly
happy to consider rotting tofu an exceptional delicacy.

It's no coincidence that both these examples are fermented
foods. Different cultures have different tastes, and the most extreme
examples are always those which a culture chooses to ferment. For
most of us, for most of our history as a species, food supply has been
inconsistent. Fermentation – essentially a kind of 'controlled decay'
using micro-organisms – is a way of preserving food. Apples rot.
Apple juice lasts longer. Leave it in the open and it will ferment into
cider, and last even longer. Distil it, and the calories in your apple
brandy will be there for you indefinitely. Our fermented foods and
drinks aren't just about what we have growing near us in different
parts of the world; they are also about the methods and traditions by
which our given culture has intervened to preserve those foods. The
flavours of fermented food and drink are in part a cultural construct.

An experiment in 2014 probed the associations different cultures
have between colours and odours[19] – for example, people who are
familiar with them match the smell of strawberries to the colour
red or pink. The researchers tested such associations across six
different cultural groups, including Dutch, Netherlands-residing-
Chinese, German, Malay, Malaysian-Chinese and US residents.

Within each of these groups, people picked consistent colours to match with each of fourteen different odours, proving that the idea of colours and flavours going together is not a random choice. Between cultures, some – such as US and German residents – had similar associations. But others were quite different. For instance, 'vegetable' is mainly lime-green for Malayan-Chinese, olive green for Malayans, and brown or yellow for the Dutch. 'Plastic' is black for Malayans, white for Dutch; while 'musty' is brown for Americans, Dutch and Malayans living in the Netherlands, orange for Malayan-Chinese, and yellow for Germans.

In evolutionary terms, we all have the same nutritional needs, and the same responses to some basic tastes. But every culture has put their own spin on what's nice to eat and drink.

Beer is a cultural as well as an agricultural product, and is surely a fascinating area for the academic study of cultural differences in terms of taste perception like the one quoted above.

Beer – technically, an alcoholic drink where the primary source of fermentable sugar is grain rather than fruit – is made and drunk in every single part of the world where grains grow, apart from states where alcohol consumption is prohibited by law.

When I travelled the beer-drinking world for my second book,[20] I learned that there are some remarkable similarities in beer across the globe. Wherever it is drunk, there's a democratic openness, a lack of pretension, which is one reason why discussions about beer such as this one sometimes meet some resistance. Beer is about quiet celebration and everyday reward. It breaks down barriers and brings people together as equals.

There's no better example of this than when, early during Barack Obama's presidency, a white cop, James Crowley, arrested a Black man, Henry Gates, for breaking and entering into a house, only to discover that Gates was a university professor who lived in the house he was supposedly breaking into. Here were all the ingredients for yet another conflagration over the American police's questionable conduct around race relations. To defuse the tension, Obama invited the two men to the White House for a beer.

If he'd invited them for a formal meeting, each of the three men would have been acutely conscious of their respective roles, status and titles. By having a beer in the White House Rose Garden, Obama was saying, let's just settle this informally, over a beer, in the way so many disputes are settled every day before they escalate into something else.[21] It worked. Later, Gates joked: 'We hit it off right from the very beginning ... when he's not arresting you, Sergeant Crowley is a really likeable guy.' The two have been friends ever since.

Beneath these universal truths, beer culture varies widely around the world. Remarkably for a product that's the third-most widely consumed beverage on the planet, there's no globally dominant beer brand like there is Coca Cola in soft drinks, or Starbucks in coffee. Every beer drinking country has its own dominant brand, so if you like a mainstream lager (and most drinkers do) you'll be drinking Bud in America, Baltika in Russia, Mythos in Greece and Tusker in Kenya. As Frank Zappa famously said in 1989, 'Every major industrialized nation has a beer (you can't be a Real Country unless you have a beer and an airline – it helps if you have some kind of a football team, or some nuclear weapons, but at the very least you need a beer).'[22]

When I visited Japan in 2004, beer's informality was, ironically, very formalised. Beer was 'the switch from on to off' at the end of the day. Before the first sip, a salaryman was in work mode. After that sip, he was free to say whatever he wanted to his section leader, for as long as beer was being drunk. A woman who went home to practise the piano after work felt she had to have a sip of beer before she was allowed to touch the keys. In Germany, I'd see people drinking bottles of Becks on the train station on their way into work, because it was only one, and they weren't going to have another, so what was the problem?. In Leuven, Belgium, elderly women sat in the train station café sipping strong Trappist beers. In Prague, beer was sold from hot dog kiosks in the middle of the main shopping streets. All of these practices would have been frowned upon in the UK. At Oktoberfest, if you weren't standing on a bench with your arms around the total strangers next to you, you weren't being sociable. Try the same thing in your local British pub, and you'll be kicked out for 'anti-social' behaviour. Cask ale is old and boring to many Brits, but cool and exciting to American craft brewers. Traditional Belgian beers are endlessly fascinating to me, but rejected by young Belgian craft beer fans.

While globalisation is inevitably ironing out some of the creases and quirks, beer remains a product that people use to help formulate an identity – as individuals, as groups, even as nations. Your cultural and ethnic background is bound to affect your approach to tasting.

Meanwhile, on a global scale, the cultural importance of music is probably best summed up by the fact that, to paraphrase Frank Zappa, you can't be a proper country without a national anthem. Most have words. But all of them have music: a song that is specifically designed to articulate what it means to be part of this nation.

One level down from that, many cultures have music that helps define them. The folk traditions of different countries can inspire the same levels of pride in their native population, and sometimes dislike or even bewilderment in others, as fermented food and drinks do. When you get off a plane in a foreign country and jump into a cab, few things make you feel as alien as the local music on the radio.

This is why every action film set in Iraq establishes the scene with a burst of Arabic music; why New York is Gershwin's 'Rhapsody in Blue'; and why the shot of a red double decker bus on Westminster Bridge with the Palace of Westminster Parliament in the background always has some stately classical music that sounds like the theme from *Yes Minister*.

Filmmakers, programme makers and advertisers use music to tell us how to feel on a continual basis. Music is ubiquitous and culturally encoded. It tells us what to expect and how to feel. It sets the scene. Different characters in a play, opera, musical or film even have their own musical themes, which play when they come on stage or on screen. Music is not just a *product* of culture and ethnicity – it creates and shapes culture on a continuous basis, every day of our lives.

Personal background

Even before you're born, your taste preferences are shaped by your individual lived experience. We each have a different genetic inheritance, which might influence anything from an intolerance to gluten to being one of the thirty per cent of the population who have a specific gene that makes coriander taste like soap. The foods your mother eats when you're in the womb have a significant effect on your early food preferences, which is why my goddaughter Holly was eating olives pretty much as soon as she could pick them up. The same thing happens with mother's milk during breastfeeding. What you're fed as a child sets preferences even further.

Then, you hit your teenage years and seek to rebel. When it was less widespread, declaring yourself to be a vegetarian at the age of thirteen was possibly the most common way in which children first started to establish their individuality and independence from their parents.

Then we start to experiment, outside the family. When I'm writing or talking about cider, about half the people I engage with say they don't like it. And every single one of them gives me the same reason: it's what they drank when they were first dabbling with alcohol as an adolescent, and they got so sick on it, they haven't been able to touch it since.

I was doing an event at Jamie Oliver's Fifteen restaurant in London, in which I matched Hoegaarden with Neil Young's 'Harvest Moon'.[23] There was one guy in the room who was sceptical about the whole thing. For him, flavour was fixed and absolute, a property in food and drink that was the same no matter who or where you were. When the song ended, I asked the audience what they thought. Most people nodded, having enjoyed the pairing. This guy was shaking his head. I asked him why, and he explained that, instead of cider, he'd once got so drunk on Hoegaarden that slightest whiff of it made him feel sick.

'Well, there you go, thank you for proving my point,' I said. 'You've had the exactly the same beer and listened to the same music as everyone else in this room, and you've had a completely different experience from them, based entirely and solely on your past experience with the beer.'

Marcel Proust's encounter with flavour and memory was so profound that it gave us 'Proustian' as a commonly used word. In the famous passage from *À la recherche du temps perdu* (*In Search of Lost Time*) he is depressed and cold, and his mother gives him a cup of tea into which he dips a madeleine, a small cake which looks like it 'had been moulded in the fluted scallop of a pilgrim's shell'. As soon as the liquid, with crumbs in it, hits his palate, 'a shudder ran through my whole body, and I stopped, intent upon the extraordinary changes that were taking place. An exquisite pleasure had invaded my senses, but individual, detached, with no suggestion of its origin.' It takes him a long time to work out where this joy is coming from. 'I was conscious that it was connected with the taste of tea and cake, but that it infinitely transcended those savours.' Eventually, he remembers that his long-dead aunt used to feed him madeleines, but the memory had lain buried for a long time. It was the taste and smell that revived it, causing him intense pleasure long before his conscious memory could catch up. 'When from a long distant past nothing subsists, after the people are dead, after the things are broken and scattered [...] the smell and taste of things remain poised a long time, like souls, ready to remind us, waiting and hoping for their moment, amid the ruins of all the rest; and bear unfaltering, in the tiny and almost impalpable drop of their essence, the vast structure of recollection.' Proust wasn't (just) tasting the tea and cake in front of him. He was tasting memory.

As we've already discussed, musical preference is also shaped from the womb onwards, and it shapes us in turn throughout our lives. My own taste in music is the result of me being a shy, love-struck teenager in the mid-1980s. Music can and will never sound the same to me as it will to a Black person born in South Central LA in 1995, or to someone from my own background born fifteen years earlier.

Intellectually, at an objective level, I can appreciate what an excellent band Led Zeppelin were – how technically proficient, how groundbreaking, and how important they are in the canon of rock music, and how so much heavy metal that came after them was a pale, lazy, pastiche of how they looked and sounded. But in my musical heart I can never love them. When I was eleven and twelve, I imagined myself to be a mod and wore a fishtail parka, whereas the overwhelming majority of kids at my school were into heavy metal, which meant that they were rockers. In their tiny lizard brains, because their older brothers had seen *Quadrophenia*, they knew that this meant they were supposed to beat up mods – identifiable by our fishtail parkas. The first thing I ever learned about Led Zeppelin was that the people who made my life a living hell on a daily basis loved them. I will forever associate Led Zeppelin with playground asphalt and getting intimate with several pairs of Doc Martens at once. You don't need to have a great deal of knowledge for such strong opinions to be formed. In 1981, I was sitting at a desk in my German class on which someone had carved the words 'Stairway to Heaven' with a compass. As a diehard Mod, I scrawled 'are shit' next to it.

The significance of any given song is just as subjective and personal as that of flavours. Within our sprawling, loose group of friends at university, there was one girl who I was in a relationship for about six months. She ended it because her mother didn't like me, which I found remarkable given that her mother had never met me, but that's the class system for you. We got on fine afterwards, and still socialised as part of the same group. Music was always playing, and every now and then, 'Eternal Flame' by the Bangles came on. Whenever it did, she'd look at me with big dewy eyes, her bottom lip trembling, and she'd remind me wistfully that this had been 'Our Song'. I never dared tell her that I couldn't remember even having heard it with her while we'd been together, let alone sharing special moments to it. I always felt like a rat, until I remembered that it was she who had dumped me. But to this day, whenever I hear it, I still see those watery eyes, and wonder how we could have had such different memories of what that song meant to us.[24]

Level of expertise or previous knowledge

Arendsnest in Amsterdam is one of the finest beer bars you'll ever encounter. At one point, it stocked every single beer being brewed in the Netherlands. There are too many to even attempt that now. There were many more than I imagined even back in 2006, when I first visited with Liz.

Between us on the table we had a bottle of Maelstrom, a beer brewed by SNAB, a brewery collective in the northern Netherlands. It was 9% ABV, so we were having one to share between us. I poured it out into two elegant stemmed glasses, and as I had recently been trained to do, took a swirl and a deep sniff. Liz followed suit, humouring my newly acquired beer sommelier skills.

Maelstrom is an English-style barley wine brewed with pungent north American hops. It's a big beer in every sense of the word, and its aromas are bold and complex. We swirled and sniffed for a few seconds, and then, simultaneously, Liz said, 'That's like sticking your head in a bag full of Parma Violets,' and I said, 'That's like walking through a pine forest.'

Objectively, Parma Violets and pine forests smell nothing alike. So how could we possibly have got such different sensations from the same beer? In my live explorations of flavour, I refer to this phenomenon as an example of 'The Estevez Conundrum' – whereby Emilio Estevez and Charlie Sheen both look exactly like their father, Martin Sheen, while looking nothing like each other.[25]

Fortunately, neuroscience has the answer. For the past fifteen or twenty years, neuroscientists have had the ability to conduct detailed brain scans. We already knew that different regions of the brain are responsible for different functions and activities. When we're using a particular part of the brain, more blood flows to that region. Because blood is warm, when this happens that bit of the brain gets hotter. So, if we have thermal image-scanning of the brain that can work in 3D, we can plot and record what parts of the brain 'light up' in response to different stimuli.

This was explored in a 2008 experiment,[26] in which jazz musicians were hooked up to functional MRI scanners while they were playing two different styes of music. First, they were asked to play what the researchers described as 'over-learned' music – basic scales that had been played by rote a thousand times. Then, they were invited to improvise freely. The heat maps of the same brains while playing the two styles are quite different. When improvising, the prefrontal cortex, which acts as a 'mental sketch pad', guiding intelligent thought, action, and emotion, and inhibiting inappropriate thoughts or distractions, was largely deactivated. But it was used playing learned scales. This research is still at a very early stage. But scientists are hopeful that, as it develops, we may finally find a cure for jazz.

If Liz and I had been hooked up to scanners like these while we were drinking our Maelstrom, our brains would have looked quite different. When we encounter new information, humans look for patterns, and we try to file things away neatly – if you see a species of bird for the first time ever, you'll file it away next to whatever you already know about birds, noting the differences as well as the similarities. If you'd never seen a bird before, you wouldn't have much of a mental filing cabinet to put it in, and you'd have to conceptualise it more broadly. Maybe you'd file it away next to other things you'd seen with wings – which means you might think of it as a tiny feathery airplane, or a terrifyingly massive fly, depending on your experience. Lucky that you know about birds, really.

Brain scans on sommeliers show that when they taste wine (because, obviously, this is where the experiments have been done, rather than with beer tasters) their brains react differently from those of novices. Experts show greater than expected activity in the region where taste and smell inputs converge, obviously. This gives them a more vivid representation of flavour than the average person. But on top of that, their experience is also more analytical. For them, tasting is also an intellectual experience, using conceptual understanding and language. Flavour training is about changing the brain's ability to interpret the signals it receives from the oral and nasal cavities, rather than 'improving' the palate – you can't grow more taste buds just by practising.

When Liz and I were drinking Maelstrom, the most important difference between us was that I had recently been taught how to taste beer properly. Not only had I learned how to swirl and sniff and all that, I'd also been taught what kinds of tastes and aromas to expect, what to look for (with your nose and mouth rather than your eyes) depending on what you're drinking. I had been *primed*. If you know nothing about beer, and you go to a tutored tasting and you're asked out of the blue what you can smell, you're probably going to struggle to do much better at first than say, 'It smells like beer.' But if the tutor then explains that this is a barrel-aged imperial stout, and that this style is often characterised by chocolate, coffee grounds, vanilla, and hints of tobacco, it's highly likely you'll suddenly be able to pick up these notes in the beer when you couldn't before. If you remember your lesson, next time you're drinking an imperial stout, you'll raise it to your nose with a predisposition to look for coffee grounds and tobacco, and if it's a decent example of the style, you'll find them. Recent research indicates that there's more to this than simply following the suggestions of the expert in charge. Our ability to perceive a flavour is linked to whether or not we already know a word for it.

When we tasted Maelstrom, I knew that high doses of American hops had notes of citrus, pine resin and cattiness. Of these, pine resin is the characteristic I like most – I find it romantic and exciting, evoking memories of silent walks by Scottish lochs and fallen needles in half-empty selection boxes. Also, it smells pretty good regardless of that. I raised the beer to my nose, expecting, hoping, to find these notes in there, and sure enough, I did, because they were there in spades. If I'd been wired up to the brain gizmo, you'd have seen the area of the brain associated with pleasure light up. But you'd also see a lot of activity in the bits associated with learning and memory. I was using recently acquired intellectual knowledge as well as sensory pleasure to get to pine forests. I was also being a bit of a pretentious knob, which seems to come with the territory – I could easily have just said it reminded me of pine resin rather than poncing about with walking through forests.[27]

Liz enjoyed the beer just as much as I did. After all, she has excellent taste in most things, except for her fondness for the band Erasure. But she neither knew nor cared about the various flavour characteristics of different hops. Still doesn't. With the scanners on, you would have also have seen her pleasure receptors light up, just like me. But there would have been no activity at all in recently acquired knowledge – she didn't have anything in there about beer. Her mental filing system had to go far and wide in search of something alongside which to classify this incredible flavour experience. While I was classifying my first eagle next to an albatross, she was thinking giant feathery bluebottles. It's not that she'd never smelled a pine forest before, but without the priming, this was more like a Proustian moment. She was jumping to flavours that filled her with a similar sense of delight, and she had to go a lot further back in her memory, to childhood, to find them. For her, travelling down different neural pathways from me, Parma Violet was a closer association than pine forest.

Take knowledge far enough and you get to the level of connoisseurship, where the appreciation of flavour can be predominantly an intellectual experience. (In your FACE, Socrates.) On the few occasions when I've drunk an 1869 Ratcliff ale, my enjoyment was at least as much from knowing the story of the beer, and knowing that I was drinking something that's over 150 years old, as it was from the character of the beer itself. Meanwhile, James May, who knows nothing about beer but still got to present a TV series about it, described the same beer as tasting like 'Magwitch's underpants'. Sometimes, ignorance is not bliss: it's just plain old-fashioned ignorance.

In music, how much you know can also have a seismic effect on how much you enjoy. Music snobbery comes in many forms, and I can be guilty of it myself. But – look, sometimes it's simply elitist. I just don't *get* opera, OK? If the story has to be explained to you, the work onstage isn't doing its job. To my ear, it all sounds the same, and I don't find the sound of operatic singing to be appealing. I find it over-acted, over-sung, and overwrought. It's grating, monotonous and shrill.

But I strongly suspect that if an opera-loving friend sat me down for an afternoon and talked me through it, playing a few pieces and pointing out what I should be looking – sorry, doing it again – *listening* for, I would learn to understand what it's trying to do. And if I did that, I imagine I would start to like it.

I have a few friends who are, or were, professional music writers. I'm always struck by the fact that even when we share the same few bands as our favourites, every one of them has a musical palate – not just knowledge, but a genuine liking – that is far broader than mine. Mine, in turn, is broader than most other people I know, because I'm the only one who has read music papers and magazines since I was thirteen. Researching this book, and growing my knowledge about music, about why it happened and where it fits in, has broadened my taste even further, and for that, any reader should be grateful.

Perceived expectations

A friend of mine is a very enthusiastic wine drinker. She occasionally visits a trendy wine shop in London's Broadway Market. She follows staff recommendations there, and one time came away with a bottle of white wine that had cost her £25. She took it home and opened it. Immediately, she was assaulted by a musty aroma. The flavour was full-on, and way too tannic and harsh for a white wine. She assumed it must have had some kind of infection, and poured it down the sink.

The next time she was in the shop, she was recommended the same wine. She explained what had happened last time: the wine was unpleasant, surely infected, and she'd had to pour it away. The assistant explained that this was orange wine. Unlike normal white wine, it stayed in contact with the grape skins for a long time. This gave it these full-bodied, tannic characteristics. It was meant to taste like that. Also, between the lines, the assistant managed to communicate that orange wine was incredibly fashionable among East London Millennials right now. My friend duly paid another £25 for the same wine she had earlier poured down the sink, took it home, and enjoyed it a great deal.

We know what we like and we like what we know. Even those of us who enjoy trying new flavours are hard-wired with preferences and expectations.

But our tastes can be manipulated. Heston Blumenthal writes about an experiment that explored this phenomenon in laboratory conditions, and subsequently inspired him to rethink how he wrote his menus at the Fat Duck.[28] In the experiment, different groups of participants were given an identical dish. Some were told it was cold smoked salmon mousse. Others were told it was smoked salmon ice cream. The first group evaluated the dish much more positively than the second group, because the concept was familiar and fitted with their frame of reference, even if they hadn't necessarily had smoked salmon mousse before.

Blumenthal later did a similar thing with a core dish in his menu. Snail porridge? You can stick that where the sun don't shine. Escargot risotto? Why, get that in my face right now.

I witnessed a similar phenomenon when I worked on the advertising for beer back in the late 1990s. We assembled a focus group of young, professional, affluent beer drinkers. We showed them a mood board full of images of people in suits running for trains while on their mobile phones, eating in fine restaurants, and playing squash. There were pictures of expensive watches and designer label clothes. As was the intention, the respondents identified with this image. This was who they wanted to be. Next, we told them that we'd launched a new beer aimed at these people. We brought out samples, pre-poured into blue glasses, so the colour of the beer couldn't be seen, and got them to taste. Everyone loved it. This was a new style of beer – they'd never come across anything like it before. Was it one of those Belgian beers they'd been hearing about? It was different from the mainstream, more interesting, just like they were! Where could they find it? They couldn't wait to buy some. So we told them that the beer was in fact Wadworth 6X, a cask ale brewed in Wiltshire. By the end of the group, no one liked the beer any more. It just wasn't for them.

Our expectations of music are also guided by bias and expectation. I adore New Order, so if you play me a new single by them, unless it's really dreadful (and there have been one or two), I'm going to love it. I don't like heavy metal, so if you do, I've already decided I'm probably not going to like it before you even play it for me.

There was a kid in my class at school – let's call him Dave, because most kids at school were called Dave, even though Dave wasn't, really – who got rumbled on this quite deliciously. It was 1982, and the Steve Miller Band were riding high in the charts with 'Abracadabra'. The chorus was one of those that you sang in the corridor without even realising it. Dave was one of those guys using the strategy later perfected by Richard Hammond back in his *Top Gear* days: hang around with the bullies, cheer them on, and that way they probably won't pick on you.

Until one day, they did.

'Hey Dave, do you like the Steve Miller Band?'

Dave had seen this before. He knew it was one of those questions that could be a trick. Especially if you were one of the few people who hadn't heard the Steve Miller Band. Dave tried to be clever and keep his cover intact.

'Are they heavy metal?'

'Yes, Dave, they are heavy metal.'

'Yeah, I love 'em.'

'Only kidding, you twat. Of course they're not heavy metal.'

'Yeah I know. That's why I hate 'em, really.'

'What do you mean, you hate 'em? They're heavy metal, Dave.'

'Yeah, aha, I was kidding as well. I love 'em.'

Once you're caught in a spiral like this, the instigators can carry on turning the screw as long as they want. And they did.

I find it refreshing, as well as slightly scary, when I hear a piece of music for which I have no context, no previous associations at all. There's a song in Kenya that, when you visit as a tourist, you hear absolutely everywhere. It's called 'Jambo Bwana', and was a hit for a band called Them Mushrooms in 1982. It was later covered by seemingly every band in West Africa. The first time I visited Kenya, I couldn't have filed it or contextualised it if my life depended on it. Was this the Kenyan answer to U2, Pulp or Jive Bunny? Was it cool or not? It wasn't heavy metal, and I wasn't Dave. Surely I could think for myself? Eventually I decided I liked it because I was on holiday, which is the only context I could frame it in. Which brings us to …

Atmosphere and surroundings

I like to ask people: what's the best beer you've ever drunk? Here's a typical response:

'Oh, my god, I remember it so well! We were on honeymoon in Zanzibar, in Stonetown, and there's a bar there right on the harbour that goes out over the water on a wooden pontoon. When you sit right at the end, you're in the middle of the water, and the dried palm roof above the deck sort of half covers the table and half not, and the sun slowly moves round till you're sitting right in it. We didn't notice because we were sitting there with our shades on. The water was so bright, really, really blue, and the sun was sparkling off the ripples like twinkling stars. I'd never seen anything so blue, so pure. Every thirty seconds or so, a shoal of flying fish would break the surface and skip along it. We couldn't take our eyes off the water, and suddenly we realised we were baking in the sun and viciously thirsty. We ordered a couple of beers and two minutes later there was a waiter in a white starched jacket with a silver tray. There were two pilsner glasses, frosted because they'd obviously just come out of the freezer, and two bottles of beer with chunks of ice sliding slowly down the sides. The waiter poured the beer and it had a thick foam on top of the gold, and we. Just. Necked. Them. Best beer I have ever, ever had.'

'That sounds amazing. What was the beer?'

'I have absolutely no idea. The local beer. Doesn't matter.'

'What did it taste like? Was it particularly bitter? Clean? Watery? Strong?'

'Sorry mate, can't remember.'

Yours might have involved an infinity pool, a mountain view, a long, parched walk beforehand, or a wedding reception. But we all have one, the same way we all have Dorada, or an equivalent holiday beer, that just didn't taste anything like the same when we brought it back home.

Charles Spence calls this the Provençal Rosé Paradox, because it happens in wine, too.[29] Professor Barry Smith, a 'self-styled philosopher of wine', tells pretty much the same anecdote as the Zanzibar one above, but it's on the Côte d'Azur, eating delicious sea food with a loved one. The Provençal rosé is 'one of the most enjoyable wines you've ever had', so you buy a case. Back home, 'it has lost all its savour… it is not even that enjoyable.'

It works the other way around too, though this is less remarked upon. I was once on holiday on Spain's Costa Blanca. Somehow, in a small coastal resort, right on the beach, was a bar that boasted almost a full complement of Belgian Trappist ales. At least five of these are in my top twenty beers of all time (unranked, before you ask.) I had a Westmalle Tripel. Served ice-cold, in 35-degree heat, on blinding white sand as the azure waves crept in, it was utterly disgusting. I switched back to a lovely pint of Cruzcampo for the next round.

In a more everyday sense, the same principle applies to how the décor, ambience and sounds of a restaurant, pub or bar influence our perceptions of flavour. In food, this has been proven many times over. In one experiment, the same meal was served in a variety of different locations, including a cafeteria, a military canteen, a grill room and a science lab. Perceived enjoyment among the diners varied by up to ten per cent.[30]

Music is absolutely fundamental to this. Classical music makes people spend more in wine shops than pop music. In restaurants, louder, faster music makes people eat quicker, which is great if you're busy. Loud music has also been shown to 'block' some taste perceptions. But slower, quieter music makes people eat more slowly, and they spend up to twenty per cent more on drinks.

Music marks the passage of time. Slower music makes people linger longer. If you've ever been in a bar or café that has a short playlist running on a loop, as you notice songs second time, you start to wonder if you've been there too long. Hear it come round for a third time, and there's a strong feeling that you've outstayed your welcome.

I'd like to invite every pub landlord and craft beer bar manager reading this to think long and hard, and dive into this research more deeply, before simply allowing your on-duty staff to unleash their favourite Spotify playlists at full volume onto a mostly empty bar.

Music can also actively change people's buying preferences. In one 1997 experiment,[31] supermarket shelves were arrayed with four French and four German wines at comparable price points. A speaker on the top shelf alternated daily between playing French accordion music and German bierkeller music. The wines were swapped around halfway through, to eliminate any bias based on their position.

When the French music was playing, seventy-seven per cent of the wine sold was French, compared to twenty-three per cent German. When the bierkeller music was playing, seventy-three per cent of the wine sold was German, compared to twenty-seven per cent French. The shoppers were interviewed about their purchases, and eighty-six per cent of them said the music had had no influence on them. The evidence suggests otherwise.

One interesting quirk in the data, which the researchers chose not to expand on, is that on the German days, overall wine sales of French and German combined were down by thirty per cent compared to the French days. People might have been buying more German wine than French, but they were buying less wine overall. Is this because the bierkeller music was prompting them to buy beer instead of wine? That seems obvious to me, but the researchers didn't record it, so we'll never know.

Music then, creates atmosphere. But it's also profoundly influenced by atmosphere and context itself. David Byrne, former lead singer of Talking Heads, has a compelling theory that music is actually determined by its context, that we 'unconsciously and instinctively make work to fit pre-existing formats'.[32] Legendary New York club CBGBs, where Talking Heads first found their own sound, had walls, furniture and a bar that create great sound absorption and uneven acoustic reflections, with no reverberation. The crowd was noisy. All this suited loud, angular music with the energy and scratchiness of post-punk and new wave bands, and those were the kinds of bands that prospered there – Television, the Patti Smith Group, Blondie, the Ramones and the Police all shared sonic characteristics with Talking Heads that worked particularly well in that venue. Meanwhile, heavily percussive music works very well in open spaces. Intricate and layered rhythms work very well on the African savannah, and that style characterises a lot of African music.

According to Byrne, this music would 'turn into a sonic mush in a cathedral,' but the long reverberation time and big, echoey stone buildings works perfectly for organs and choirs. Mozart began playing in palatial drawing rooms – big rooms, but not as big as concert theatres – full of people in fine costumes. The sound absorption and intimacy meant his intricate stylings were crystal clear. The composition of classical orchestras had to change when they moved from these rooms into large theatres. Today, most of us can tell when a band has their eyes on playing stadium gigs. U2, Simple Minds and Coldplay changed their sound specifically to suit the massive arenas they aspired to playing, with lots of big statement chords, and wide spaces between them to allow time to reach the back.

How are you feeling?

If you're happier or more relaxed, if you're comfortable and warm, if you have less noise going on in your head, there's simply more capacity to truly experience what you're tasting, and a more favourable environment in which to appreciate it. If you're upset or stressed, you might drink more quickly and not notice the flavour as much. Meanwhile, if you like the music that's playing, this can put you in a better mood, and that can make you more susceptible to enjoying the flavour of what you're drinking.

As we've seen, music creates, enhances and changes our emotions. It's all about feeling. This is why we create different-themed playlists. The compilations we put together for the gym, and for working are quite different. I have one called 'Cooking' and one called 'Dinner', which have quite different moods. Going back to movie music and how we feel, the theme tune to *Jaws* is now almost a genre of its own. Creatives steal it and use it as a shorthand when they want you to feel tense and excited, to breathe heavier, because you know something thrilling is about happen. Put that same theme on your sexy bedtime playlist, and it really doesn't have the same effect. Trust me.

Priming and presentation

A clever server can do a lot more in the presentation of a drink to heighten your expectations, and, therefore, your perception of how much you enjoy it. Now that branded glassware for big brands is the norm, if you order a pint of Guinness and a pint of Stella, they are simply not going to taste as good if they're served in each other's glassware as they will if served in the right glass. There is some truth about the science of the shape and thickness of glassware bringing out certain taste notes in a beer. But it's comparable to how hi-fi buffs geek out about gold speaker cables. It does make a difference, but you'd have to be extraordinarily gifted to pick out that difference with everything else that's going on. Branded glasses are more about making the beer look nice, and making you feeling like you have purchased a premium product. Guinness goes one better: the two-part pour heightens your anticipation, which is going to make it seem to taste even better, unless there's something seriously wrong with it.

I haven't seen an experiment where people are served vintage Champagne out of plastic cups and cheap Prosecco from crystal flutes, or Westmalle Tripel in a cup from the top of a Thermos flask and Carling from a Belgian chalice. I don't need to. I know exactly what the results would be. When the late Charles Campion and I hosted a beer and food dinner at the restaurant at the Cadogan Hotel in Knightsbridge, we served Aspall Premier Cru in champagne flutes as a welcome drink. A friend of ours, who only ever drinks expensive wine or gin (not the same person as in the orange wine story above) grabbed one as she entered and sank it quickly. A waiter appeared at her shoulder and said, 'Can I top you up with cider, madam?'

'Oh no,' she said witheringly, '*I'm* drinking the *champagne*.'

'No you aren't, madam,' said the waiter, holding the bottle towards her so she could see. Her face was a picture. And not a happy one.

Not all these effects work in the same way at the same time. Sometimes they might contradict each other. I wrote above about how if you're primed at a tutored tasting, you might be able to get those flavours again next time you try the same drink. This doesn't always happen. Early in my career, I attended several beer and food tasting sessions hosted by Garrett Oliver, the brewmaster at Brooklyn Brewery and an all-round gourmand. As well as being an incredible brewer, Garrett is a talented and compelling presenter. He weaves stories around each of his beers. When you're with him, and he's talking you though them, it's as if he's taking you inside the beer itself. Everything he says, every nuance of flavour he pulls out, resonates. You want to please him by detecting every hint and note he mentions.

My tasting notes from those sessions were always full of purple prose – and they weren't just the words Garrett had used. Several times, I'd take bottles of the beer home with me. In my living room, I was caught between the cognitive priming of the beer educator, and the environmental influence of the Provencal rosé/Dorada effect. I would sit there saying to myself, right, I know what's in this. I can still remember what it tasted like. These are my own words, right here. I am now going to evaluate thus beer again, with my professional taster's head on. While the home beers were never less than great, I never got a fraction of the flavour experience I'd recorded while being in the room with Garrett. That experience hadn't just been about the information he was imparting to us. Maybe it was his charisma and presentation technique. Maybe it was because he had brewed the beers we were tasting. Maybe it was feeling special for having been invited to such an exclusive event. But more than once, being in the room with Garrett made his beers taste better than when he wasn't there.

Of course, I've had many musical Garretts in my life. If you're into music, so have you. I was predisposed to like anything I heard

John Peel play on his late-night radio show, simply because it was him playing it. I've seen research that says the main reason people begin smoking in their teens is because there's an older, cooler kid who smokes, and they want to be more like them. I never smoked. But I studiously copied the musical tastes of several kids I wanted to be more like.

Need a drink yet?

So, we've established, to the best of current scientific knowledge, how flavour works and how music works. We've shown that what we might think of as fixed properties in beer and music are, to say the least, open to interpretation and misinterpretation. And we've seen how pretty much everything else around you and about you can have a direct impact on your perceptions of flavour. In this, we've already started to see how music in particular can affect your perception of flavour, and your behaviour around drinking.

We've already got enough stuff here to start playing around with matching certain songs and styles of music with different flavours right now. A lot of the matches on Side Two of this book rise directly out of what we've already discussed.

But there's another layer to come, that's all about what happens inside the brain when all these different sensory stimuli hit it at once. I was going to say this is the fun part, but it's all fun. It's probably more accurate to say that this is the bit that gets really mind-blowing. Pour yourself a glass of something lovely for the next track. You might need it.

Crossmodal correspondences – how the senses interact, and why you can hear the difference between bitter and sweet

Sensory overload

Have you ever read a page in a book and then realised you can't remember a word that was on it? Or had a partner talk to you and then rumble the fact that you're not listening, asking you to repeat what they just said, and you can't?

Of course you have. We all do. Sharing these experiences is part of what we recognise as being human. Your sensory organs were doing their job, but your brain wasn't taking any notice.

As we've established, in a meaningful sense, we 'taste' not with the mouth or nose, but in the brain. Likewise, while the ear picks up information, it's the brain that makes sense if it. Most of the time we're drinking beer, or listening to music, our attention is focused on something else. The brain can't focus on and appreciate, say, the subtle balance between grassy hops and the gentle biscuity, Ovaltine character of the malt, or the modulation from major to minor in the middle eight of the song on the jukebox, if it's too busy concentrating on your mate's latest dad-joke, or whether England are going to go to penalties. It's only when we're consciously and mindfully tasting or listening that the brain is focusing properly on the information it's receiving from our nose and mouth or ears.

Each of our sensory organs is receiving information, processing it into electronic signals and sending these to the brain on a constant basis. It's pretty much impossible for the brain to pay attention to all of them at once. We automatically close off some senses to focus better on others.

But at the same time as our brain is ignoring some senses, it's trying to integrate what it's getting from others, trying to build up a complete picture. Raise an object in front of your eyes. Look at it with one eye closed. Then the other. Then both eyes open. Two eyes give you a fuller picture. In the same way, your senses combine to give the best overall impression they can of any external phenomena that grabs your attention. That looks like a large object heading towards you. Is it roaring angrily? Does it smell like an animal? The senses overlap, help each other out. You can smell fear and attraction, see speech, and hear space.

This sensory overlap is reflected in our language. If I talk about loud patterns in fabrics, brightness, texture or heaviness in music, or roughness, smoothness or sharpness in taste, you know exactly what I'm talking about, even though I'm using descriptors from the 'wrong' sense. You've probably told someone that you can see what they're saying, or that their actions smell or feel a bit dodgy.

Occasionally, and unintentionally, our senses get it slightly wrong, and confuse or deceive each other. Think about ventriloquism. 'Throwing your voice' has nothing at all to do with sound – it's all about convincing the eye that the voice is coming from a different place. When you're sitting on a train in a station waiting for it to set off, you can *feel* it beginning to move before you realise it's the train on the next platform setting off in the other direction, and you're still stationary.

All of this is the brain trying to combine the 'right' information from the 'right' senses, in a way that's appropriate to the current situation. And for the last twenty years or so, we've been able to study this much better than ever before.

The practice of trepanning – basically, drilling a hole in someone's head – goes back to Neolithic times. The Egyptians were writing about the symptoms of brain damage as far back as 1700BC, but even 3000 years after that, the notion of drilling into someone's skull to let out evil spirits that were confounding the patient's brain was believed. Even in the middle of the twentieth century, experiments

on the brain still began with going through the skull to manipulate
brain tissue and see what happens.

Now we know that neurons are cells specialised for commun-
ication. They are able to communicate with each other and with
different cell types via specialised junctions called synapses,
through which electrical or electrochemical signals can be
transmitted from one cell to another.

Traditional neuroscience was seen as a branch of biology.
Now, thanks to new brain-imaging technologies, drugs that can
manipulate neurotransmitters such as dopamine and serotonin,
and computer modelling of how neurones work, 'cognitive
neuroscience' has evolved into an interdisciplinary science that
combines chemistry, computer science, engineering, linguistics,
mathematics, medicine, philosophy, physics, psychology...
and now, drinking beer and listening to tunes.

Using these techniques, we've learned about brain plasticity:
if you lose one sense, the part of the brain that's normally linked to
that sense will help out elsewhere, taking cues from other senses.
There's now a large body of research that shows this is part of a wider
pattern of how the brain attempts to process sensory information.
We make shortcuts and automatic correlations – or more accurately,
correspondences – between different sensory stimuli. We sense
patterns in them, and deep-seated relationships between them. For
example, various experiments have consistently shown that people
consistently match high-pitched sounds with small, bright objects
that are located high up in space, and low-pitched sounds with bigger,
darker objects that are lower down. Developmental researchers
have shown that children as young as two match loud sounds with
large shapes, and vice versa. These 'crossmodal associations' have
been established between auditory pitch and smell, smells and
shapes, and even shapes and flavours.[33] It's now an established
scientific fact that our senses don't just work in isolation, with the
brain filing the input from each sense in its correct box and then
combining them. These senses overlap, merge, and bleed together.

Synaesthesia – and why this is not it

For as long as I can remember, I've perceived numerals as having different personalities. 1 is stolid and cool. It reminds me of the actor Robert Mitchum. In fact, it looks exactly like him. 6 always stands out as neurotic and highly stressed. It's baring its teeth and hissing. 4 is a bit goofy and very gregarious. 5 is friendly, always cracking a joke. It looks like it's nudging you and guffawing. Would it be OK if we just leave 8 of out this for now? That's private. Also, be warned – 9 is sly and devious, and cannot be trusted.

I used to think that everyone saw numerals in this way, that it was as universal as perceiving the sky as blue, snow as cold, or Percy Turner's from Jump, near Barnsley, as the best pork pies in the world. Until one day, I mentioned it to a group of friends, and they questioned my sanity.

It turns out this is a common experience for all synaesthetes: we all think everyone else perceives reality the same way we do, until we find out they don't.

Synaesthesia is, mostly, a condition whereby the wires between the senses and the brain have been connected in a different way. Some people smell or hear colour, taste sounds or see music. There are other sensory correlations too: a common one is that different numerals or shapes have different colours, so 'A's are always red, 'E's are blue and so on. I didn't realise my thing with numerals was called 'sequence personality synaesthesia' until I read about it in one of the papers I used for this research. It was nice to learn that I'm not the only one, even though it's a rare variety. Other kinds of synaesthesia help people become more interesting artists or composers. Mine's useless in that regard. It hasn't even been any help with maths. But at least I get to laugh at 5's jokes while you don't.

Unsurprisingly, given how unusual it is, not much is known about the nature of synaesthesia, except that it definitely exists. I've seen studies that say it affects one in 23 people, and others that say it's one in 21,000. Because so many people with it assume everyone else perceives things the same way they do, it's impossible to measure unless someone realises they have it in the first place. Some researchers say there's a scale of synaesthesia – we're all on it somewhere, but some of us are much further along it than others. Others suggest that synaesthetes are totally separate from other humans, like the mutants in X-Men. Some say synaesthesia can be induced among non-synaesthetes by using psychedelic drugs. This would make a lot of sense in the context of the lyrics and music of the late 1960s and early 1970s. But I'm guessing further exploration of this theory today within mainstream academia might run into a few issues.

Synaesthesia and crossmodal correspondences obviously look very similar at first glance – both are about sensory information getting mixed up. At the start of the twenty-first century, some of the experiments starting to look at crossmodal perception were motivated by a hypothesis that it's a weak form of synaesthesia; back to that idea that we're on all the scale somewhere. But Spence and others refute this. Whatever synaesthesia is, it manifests as something that's highly unusual and quite different from how most humans perceive the world. Spence argues that, in contrast, crossmodal correspondences are meaningful because they are shared by many people, and may even be universal.[34]

Sound and vision

The hierarchy and snobbery around the senses that's been established by philosophers over millennia still holds strong today, which is why most of the research into crossmodal perception is heavily biased towards the relationship between vision and sound.

Seeing and hearing are profoundly interlinked. Think of lipreading. Most of us might say we can't do it – not without any sound at all. But vision contributes a huge amount to what we think we're 'hearing'. If you're in a loud pub or bar, and you're listening to someone speak across the table, just see what happens if they hide their mouth behind their hand while speaking at the same volume – or even louder.

Possibly the first ever experiment[35] in trying to establish crossmodal correspondences was carried out in the 1920s. Nonsense words, with no meaning in any language, were given to people who were asked to pair them with shapes. For the purposes of the experiment, one of these shapes is called 'Takete' and the other is 'Maluma'. Which one do you think is which?

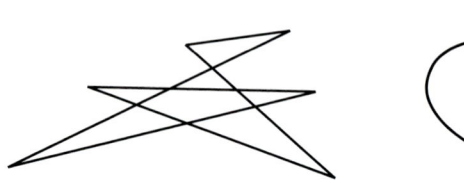 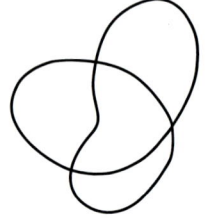

Almost everyone says the spiky shape is Takete and the round one is Maluma. It just seems obvious, doesn't it? But why? It's more than a random guess. Well, 'Takete' has lots of hard consonants that correspond with the spiky shape, while 'Maluma' sounds round. (Wait – so sound has a shape?) Another hypothesis is that when you say the words, your mouth is making round shapes to pronounce the long 'U', while there's lots of tappy tongue against the teeth with all the 'T's.

This pairing holds true across many, though not all, cultures and languages. More recent studies have shown that the association is just as strong whether you show someone articulating the words or not.[36]

There have been a great many experiments since this one was first done. Across many different approaches, researchers have consistently concluded that connections that are not random or arbitrary exist between a variety of different audio and visual features, both in terms of simple aspects such as loudness and brightness, and more complex phenomena such as shapes or images and words or short musical clips.

Flavour and vision

We've all heard about wine experts being tricked into thinking they're drinking vintage Burgundy by someone adding food colouring to cheap plonk. But this doesn't mean they've been rumbled as frauds – they've been trained to use *all their senses* when evaluating something. Mess with that training, and the results are bound to be off. Anyway, you're no better. If I give you three glasses of fizzy sweet liquid – one hazy yellow, one brown and one red – you *will* taste lemonade, cola and cherryade, even though all three glasses contain exactly the same flavour compounds.

When I worked in advertising, we always used to say 'The first bite (or sip) is with the eye.' I worked on the Stella Artois 'Reassuringly Expensive' TV campaign. In the first of those ads, our hero – a Provençal flower seller we called 'Jacques de Florette' – barters a bunch of his freshly picked red carnations for a baguette. Then he sees a pint of Stella being poured. The gag at the end is that he barters his entire cartload of carnations for the beer. Part of the appeal of the ad is the drinking shot. You can see he's hot and tired. His eyes close as he raises the glass to his lips. It's the most gorgeous-looking beer in the world (Stella was, in fact, a decent lager thirty years ago) and the viewer becomes Jacques as he was in the first part of the ad – you'd give anything to get your hands on that pint. Try as we might, we never managed to get that drinking shot just quite as perfect in any subsequent ad.

There's a reason we say that we taste with all the senses. The sight of food can make our mouths water. Think of the apple, as a mythical symbol of desire. Then think how quickly desire turns to disgust if you bite into it and see a maggot. Or how much worse, as the old joke goes, to see half a maggot.

This relationship goes to a deep crossmodal level. Before I started any of this research, I always used to say that my first sip of American IPA, and my first taste of a proper Indian curry, was like tasting in colour for the first time, and realising I'd been tasting in black and white until then. It was just a metaphor. But metaphors work because we can understand the relationship between the word or phrase and the concept it signifies.

Experiments have shown that a meal served on a square black plate or a piece of slate will taste more bitter than it would on a neutral plate. Serve it on a red, round plate, and it will taste sweeter. Changing the colour, or even the shape, of food can alter our perception of its flavour and also our impression of how much we (think) we enjoy it.

Some of this blurs over into the stuff we covered in the track about context. The distinction is hazy. Contextual factors are present here, but there's something else too. In 2014, drinks giant Diageo commissioned research to help them design packaging for luxury whisky brands.[37] They found that if you drink a glass of single malt in a room carpeted with real grass, accompanied by the sound of a lawnmower and birds chirping, all bathed in green light, the whisky tastes 'grassier'. Replace that with red lighting, curved and bulbous edges and tinkling bells, and the drink tastes sweeter. Meanwhile, creaking floorboards, the sounds of a crackling fire and a double bass playing in the background combine to bring out the woody notes and give you the most pleasurable whisky experience.

The visual cues in this experiment – the use of lighting – were playing a key part in altering how the whisky 'tasted'. But there were auditory cues there as well …

Flavour and sound

I'm sure even the most sceptical reader would accept by now, knowing everything we know about flavour, that what we see has an impact on what we taste. But what about what we hear? Somehow, that seems less likely. Not to me – but I encounter a lot of disbelief when I first try this idea on people.

It's hard to deny the appetite appeal of certain sounds – we taste with all the senses, remember – the sizzle of a steak or the cauldron bubble of chips being plunged into a deep-fat fryer. The pop of a cork being pulled from a bottle, or the hiss of the beer o'clock can being opened.

But here, we're talking about more than that. We're talking about close, non-random relationships between certain properties of flavour and certain properties of sound.

Musicians often use flavour words to describe the properties of different instruments – Berlioz wrote about the 'small sweet acid voice' of the oboe. *Dolce,* Italian for sweet, is an actual classical musical notation, often simply translated as 'sweet', as if it doesn't need any explanation. It actually means to play with a soft, light touch, to evoke tenderness or adoration. Why does that mean sweet? Well, it just does. Do you disagree?

Ludwig van Beethoven went further. An enthusiastic consumer of wine, he declared, 'Music is the wine which inspires one to new generative processes, and I am Bacchus who presses out this glorious wine for mankind and makes them spiritually drunken.'

Again, the use of metaphor here indicates the existence of a deeper relationship that allows that metaphor to make sense. Here's a quick example, without thinking: remember Takete and Maluma? Which one of these is bitter, and which is sweet? Yep, I thought so too.

The research into the relationship between sound and taste started relatively simply and has moved onto more complex and ambitious hypotheses. It's worth following the story of how it evolved – and that story will be really easy to follow if you've read the track on how sound and music works.

Flavour and pitch

The earliest experiments to explore the relationship between sound and flavour were about correlating basic tastes with musical pitch. And the very first one happened not with wine – like the vast majority of experiments that followed it – but beer.

In 1968,[38] and again in 1976,[39] Danish psychologist Kristian Holt-Hansen published the results of an experiment where he took two beers – Carlsberg Pilsner and the much stronger Carlsberg Elephant beer – and asked respondents to play around with the frequency of a musical tone until they got the pitch that felt right for the beer. The stronger beer was matched consistently with a pitch between 640–670 Hz, while the weaker pilsner was consistently matched to a lower frequency, between 510–520 Hz. Intriguingly, some respondents became quite enthusiastic about the idea of the perceived pitch and the flavour of the beer being 'in harmony' with each other. One respondent in the 1976 study gushed, 'My right hand with the glass of beer in it trembled so violently that I was suddenly afraid of dropping the glass. I felt as if I was floating in the air. The tone was intensified to such a degree that it sounded like a symphony orchestra and the room was filled with it. My jaws were moving in and out with the rhythm of the tone.'

Subsequent researchers have been unable to reproduce these dramatic effects. Probably because it wasn't the late sixties or seventies any more. They have, however, found consistent relationships between different flavours and sonic pitch.

A 2011 experiment used a selection of aromas from a kit designed to teach people how to taste wine (obviously).[40] After sniffing each aroma, they were presented with a range of musical samples varying in pitch, and asked to choose the corresponding one. They could play as many of the fifty-two different samples (thirteen notes on each of four instruments) as they wanted before making their choice.

Fruity aromas such as pineapple, raspberry, lemon and apple were consistently matched as high-pitched. Smoke, musk, dark chocolate and woody notes were low-pitched, with caramel, mushroom and vanilla in the middle.

> ### Experiment: The sonic pitch of flavour
>
> Do different flavours make different sounds? Of course not, that would be weird. But just for fun, try ranking the following flavours in order of sonic pitch, from low to high:
>
> - Vanilla
> - Lemon
> - Tobacco
> - Apple
> - Chocolate
>
> *If you ranked them roughly like this: Tobacco, Chocolate, Vanilla, Apple, Lemon, you did the same as around 95% of people who've been asked to do the experiment. For some reason, we have an innate understanding of sensory properties that shouldn't really exist if crossmodal perception wasn't a thing.*

This fitted with previous research that suggested the simple tastes of sweet and sour were high-pitched.

In 2016, a new study using beer(!) was conducted to try and update and explain those early experiments by Holt-Hansen.[41] Was the pitch corresponding with flavour or alcohol content? (Elephant is quite a bit sweeter than Carlsberg Pilsner, as well as being stronger.)

This time, the lucky respondents were given samples of three beers by Brussels brewer De la Senne: the wonderful Taras Boulba, a dry, spicy pale ale at 4.5% ABV; Zinnebir, a golden ale at 6% ABV; and Jambe de Bois, a now sadly discontinued tripel, at 8% ABV. (While we know what the beers were, respondents tasted them blind.)

The beers clearly varied in alcohol. While no one with a working palate would say they all tasted the same, they were all quite dry. The results showed that they all corresponded with roughly the same pitch: around 323 Hz, chosen from a reasonably narrow band that was offered – 50 to 500 Hz.

Then, the experiment was repeated, but with Zinnebir replaced by a new beer – Belle Vue Traditional Kriek Lambic, which is not a traditional lambic at all, because it's excessively sugary sweet with no sourness. Respondents were given a wider range of pitch to choose from this time (50 to 1500 Hz). Here, Taras Boulba was pitched pretty much exactly where it was in the first experiment. Jambe de Bois was pitched somewhat higher than it had been, and the 'lambic' was significantly higher, at 487 Hz – twice as high as Taras Boulba.

Every experiment of similar design gets the same results, and when I quiz my audience at events, there's no variation there, either. Sweetness and acidity, and the flavours we associate with them, are higher pitched. Bitterness and umami are always lower pitched. Saltiness comes somewhere in between.

Taste, timbre, tempo and volume

There are other basic mappings between tastes and musical properties, some of which feel as intuitive as the relationship between flavour and pitch. Think about tempo, or speed – are lemons slow or fast? We tend to correlate slow, heavy, dark, loud and dense with bitter; and fast, light, bright and softer with sweet. These relationships feel fairly obvious, and are possibly related to the physical properties of objects.

It gets more interesting when we look at timbre, the 'voice' of different instruments.

In their experiments exploring the relationship between pitch and taste or aroma, Anne-Sylvie Crisinel and Charles Spence also looked at different instruments, using the same range of tones on piano, strings, woodwind and brass.[42] In the first experiment, different flavour compounds were diluted in water and served in small measures. To deliver the basic tastes, simple compounds were used: caffeine (bitter), citric acid (sour), sucrose (sweet), sodium chloride (salt), and monosodium glutamate (umami). In addition, some more complex flavours were added: almond, coffee, lemon, orange flower, peppermint, rose and vanilla.

The piano is consistently seen as being sweeter than other instruments. For some strange reason, brass corresponds not just with bitterness, but a whole range of other flavours that people consider to be unpleasant – it also scored highest for acidic and for salty, although all the salty scores were lower than the other basic tastes. The results for more complex flavours, as opposed to basic tastes, weren't as distinct, but peppermint corresponds best with the piano. When respondents tasted actual coffee, as opposed to a simple caffeine solution, brass was associated with it much less strongly, while strings and woodwind compete for the strongest association – possibly because coffee is a lot more pleasant to drink than a bitter, watery caffeine solution.

More work is needed on the relationship between timbre and flavour. What's been established so far is consistent, but fairly basic. Sweetness is loved, bitterness – among the population as a whole, rather than beer drinkers – is not. The piano wins in any experiment as the sweetest sound, possibly because it sounds more agreeable than other instruments when it's played on its own in a laboratory setting. We listen to piano solos all the time. Brass on its own – if you didn't grow up surrounded by the music of colliery brass bands as I did – can be a little more challenging, I guess. It would be interesting to see similar experiments done with acoustic and electric guitar and different synth sounds.

Basic tastes and musical styles

Argentinian Professor Bruno Mesz, a Master in Mathematics, and his colleagues created an experiment which pushed these correspondences further. Rather than simply explore individual, specific properties of music, they progressed to fully fleshed-out musical styles, with research that has become a core foundation for my own beer and music pairing.

In an ingenious double experiment, first the team gathered nine musicians and gave them an array of different words to improvise to. Each of these musicians had been playing for at least ten years, and they spanned classical, experimental and popular music styles. The words included 'sweet', 'salty', 'sour', and 'bitter', but so as not to give the game away – and also to establish a useful control – other words such as 'ferocious', 'sorrowful', and 'delicate' were included. The musicians were asked, one word at a time, and separately, to improvise music around each word.

The results for the control words were as you might imagine. There was a consensus of musical styles across the different musicians for words like 'delicate'. Incredibly, there was also a high degree of consensus on the improvisations around the taste words. Each one prompted improvisations that were consistent with each other, and distinctive from the other taste words.

But this was only half the experiment. The team then took clips of each improvisation, and played them to a range of people who had no musical training. They were then asked to match each piece of music with one of the four corresponding taste words.

Experiment: Basic tastes and musical styles

Let's pretend you're at one of my beer and music matching events!
Below are links to four pieces of music. Have a listen to the first ten to fifteen seconds of each song (except the first one – cue that up to about fifty-five seconds in – there's an unnecessary intro).

A https://open.spotify.com/track/6JW1soaXabex4P3gzzrpcC?si=9c8a3be78c8e413b
B https://open.spotify.com/track/7MeOvOSlJfaPY7Pc4Geltd?si=25caef93ca104ca8
C https://open.spotify.com/track/0WSTUuw7gP80yTcEkL7oXc?si=8f1000d376dc46fc
D https://open.spotify.com/track/2Vqi1Si1dQjMtAZ79UreqL?si=6f27a65ab7df4859

Which of these sounds sweet, sour, salty and bitter?
If you have no opinion, record it as that.

The two studies were consistent with each other. The taste words that the musicians had improvised to were chosen as the same taste words by the non-musicians who listened to them. If you want to do my experiment in the box above (p.93), try it now before reading on.

If you enjoyed that, welcome to my musical world. I chose those four pieces from my music collection, back before I expanded this to cover other people's tastes. My own experiments would not pass muster as 'proper' science because I don't conduct them under laboratory conditions. Nevertheless, when I give out score sheets and tally what people say,[43] they're totally consistent with the results of Mesz's experiment, which found:

- ***Bitterness*** corresponds with music that is low in pitch and legato – that is, the notes are smooth and connected, with no gaps or silence in between. One blends into the next. This is why I chose a cello drone (A above) for my own attempts to replicate the experiment. I get forty-four per cent of my sample agreeing that this is bitter, with sourness in second place on twenty-seven per cent. Only four per cent choose sweet. And only seven per cent tick 'no opinion'.

- Mesz established that ***saltiness*** corresponds with music that has notes of short duration and high articulation, better known as staccato. It's choppy and can sound slightly brittle. I could think of no better example of this than 'Marquee Moon' by Television (B above.) In my own attempts, this one is less convincing, although I think I know why. Salty still gets the top score, with thirty-three per cent. But it's a narrow win against sweetness, at thirty per cent. The main guitar line is double-stopped, meaning two notes are played at once, creating a harmony. This in turn plays off a second guitar that's high-pitched and melodic. Together, the two guitars do create something that might correspond with sweetness (see below.) That's what you get for using one of the best songs ever written instead of pure tones, I guess. Bitterness and sourness are quite far behind, with only six per cent of people having no opinion.

- **Sourness** corresponds with music that is, according to the original experiment, of high pitch, long duration and high dissonance. The precise definition of dissonance (as opposed to consonance) is elusive, and our perception of it changes over time and between cultures. But you recognise it when you hear it. If music is about tension and release, created by notes following an order that feels 'right', dissonance is an order that feels 'wrong'. It's when notes don't follow each other in the order we expect, or when notes played together don't feel harmonious. It's the clanging piano at the end of the Beatle's 'Day in the Life', or the sharp, main two-note melody in our old friend the main 'Jaws' theme. It creates tension and unease, and that's why John Williams used it so masterfully to put you on the edge of your seat. I chose 'Birthday' by the Sugarcubes (C above) because it sounds woozy and strange and none of the notes go where they should, as if you're having your birthday in Wonderland after too many eat me/drink me potions. For me, there's also a bit of synaesthesia going on here perhaps: it sounds like overripe fruit, and makes me see lime green and acid yellow. In my show, sourness gets a healthy forty per cent score, double that of salty, in second place. Nine per cent have no opinion.

- Finally, according to the proper experiment, **sweetness** corresponds with music that is consonant rather than dissonant – it's harmonious. It's also long in duration and low in articulation, and soft rather than loud. This always puts me in mind of close vocal harmonies (male or female), and for me this works across all genres. While researching the pairings on Side Two, I spent a profoundly moving afternoon taking sweetness as a musical principle and going on a journey from the Beach Boys and sixties girl groups through South African choral bands such as Ladysmith Black Mambazo to Gregorian chants. It's also, for me, the sound of a twelve-string semi-acoustic guitar, the signature sound of C86 indie, though never better than on The Byrd's 'Eight Miles High'. For my live show experiment though, I chose the Cocteau

Twins (D above), where voice and guitar combine in chiming, honeyish sweetness. Again, this clip scored a convincing forty per cent for sweetness in my experiment, with salty a distant second on twenty-four per cent, and only seven per cent having no opinion.

Mesz's original experiment, with its clever reversal of having musicians compose to a taste word, and then listeners picking the same corresponding taste word in response to the music, is compelling enough. But I'd like to think my own semi-scientific work has added something extra. Because I get some scepticism around this whole idea, unlike any of the experiments I've looked at, I gave respondents the option of saying, 'Nah, this is rubbish. There's no relationship between these sounds and tastes.' In the formal experiments, you were forced to pick one, even if it was only guesswork. In my experiments, if people thought the whole thing was nonsense, they had the opportunity to say so. But less than ten per cent of the sample did. With my five options, if the responses were perfectly random, each taste – plus the 'no opinion' option – would have scored twenty per cent each. Three out of four of my matches were double that. This is not random. It's a relationship that those of us who were brought up in the Western musical tradition agree on, even if we don't consciously realise it's there.

Matching versus changing perception

Academics are their own worst critics. In many of the papers I've read, they don't even wait to let anyone criticise them – they jump in before the paper has even finished, picking holes in the methodology, and suggesting areas for further research that they haven't covered. The introduction to one paper written by Charles Spence was summarising previous work on the subject, and said something along the lines of, 'If the authors of this study had bothered to think a bit harder and taken more care, they'd have realised that this aspect was wrong.' One of the authors of the study he was slating was ... Charles Spence.

In the work we've looked at so far, there's a big caveat and a big question. The caveat is: up to now, people have been responding in most studies to taste *words*, rather than actually tasting. What if it was something about the sound or the context of the words used rather than the taste itself? The question is: okay, so we've established that these correspondences definitely exist – that we associate different tastes and flavours with different types or attributes of music – but is this relationship so strong that introducing or changing different musical styles can actually *change* our perception of flavour rather than just *corresponding* with it?

A whole bunch of different experiments followed, taking these comments on board. In 2011, Spence, Crisinel and others changed the sonic properties of music that was playing in the background while people were asked to eat a piece of cinder toffee, which had both sweet and bitter flavours to it. One soundtrack was designed to correspond with sweet, the other with bitter. When the soundtrack was changed, perceptions of the flavour of the toffee changed by as much as ten per cent.[44]

Jumping back in time a few years, in 2008 Professor Adrian North at Heriot-Watt University carried out the experiment that introduced me to the whole notion of matching beer and music in the first place. It wasn't written up in an academic journal until 2011, but it was so interesting it was a story on BBC News not long after it was carried out. It was an exercise in pairing music with wine. I read it, fascinated, and immediately thought: why didn't he use beer? The university he works at is home to the country's leading brewing school! Having got over my ire, and having made a note to steal his work and apply it to beer if he wasn't going to, I read that he'd moved the dial quite far on by having students taste actual wines (a Chardonnay and a Cabernet Sauvignon, both from Chile, retailing back then at around £10 a bottle) and listen to actual music. As they were doing so, he gave them four descriptions that, intriguingly, could apply to both the flavour of the wine and the style of the music. A piece of music was chosen that fit each set of words.

These were:

- Powerful and heavy – 'Carmina Burana' by Orff, or the 'Old Spice music' to a certain generation
- Subtle and refined – 'Waltz of the Flowers' from *The Nutcracker* by Tchaikovsky
- Zingy and refreshing – 'Just Can't Get Enough', a cover of the Depeche Mode hit by ironic French lounge band Nouvelle Vague
- Mellow and soft – 'Slow Breakdown' by Canadian musician Michael Brook

Each group of respondents got to listen to a different piece of music while they drank the two wines, and were asked to score the wine from 0 to 10 on each of the four suggested attributes. The first finding was that any music at all gave the wine a higher score on each attribute than drinking it in silence. (Later studies found that wine consistently tasted better with music in general than when tasted in silence.) But there was a correlation between the 'right' wine and the 'right' musical style. The Chardonnay tasted twenty-six per cent mellower and softer when 'Slow Breakdown' was playing, and forty per cent zingier and more refreshing with 'Just Can't Get Enough'. The Cabernet was twenty-five per cent mellower and softer with Michael Brook, but a whopping sixty per cent more powerful and heavy when 'Carmina Burana' was played.

Emphatically, the music 'changed' the flavour of the wine. North – again, his own harshest critic – suggested before anyone else could that maybe the pairs of words were leading the witness. They weren't strictly describing the music or the flavour – they were metaphorical, applicable to both. So maybe they were creating a 'priming' effect, conditioning people to give the ratings they did. I'd argue that the fact these pairs of words *can* be ascribed to both the wine and the beer, that the same words work for both, does in fact suggest evidence of an underlying relationship, a crossmodal affinity between completely different sensory inputs that can be described using exactly the same words.

Still – this doubt meant there was room for more experiments to be carried out between music (usually classical) and drink (almost always wine). Methodology varied, but consistent changes in music produced perceived changes in the wine, along consistent lines.

There was one more experiment using beer, so we should close this section with that. In 2016, the authors of 'The influence of sound-scapes on the perception and evaluation of beers'[45] claimed the paper was 'the first assessment of its kind made with beer'.[46] Like the only other recent study on crossmodal correspondences using beer, the beers were the same three De La Senne ales: Taras Boulba, Zinnebir and Jambe de Bois. Respondents were given the three beers to taste, and then the same three beers again, in a random order, under different conditions. The respondents didn't know the beers were the same.

Four soundtracks were chosen from different previous experiments in which they had been paired most closely with sweet, sour, bitter and salty. Each beer was tasted with the soundtrack that most closely paired with the dominant flavour in the beer, plus one other. The sweetest beer (Jambe de Bois) and the bitterest beer (Taras Boulba) were rated as significantly higher in those attributes when the 'right' soundtrack was played. Zinnebir, somewhere in the middle, didn't get a significant result. But interestingly, it was rated as significantly higher in alcohol when the bitter soundtrack was played.

I'll leave the experiments there for now. In 2023, a 'systematic review and narrative synthesis' of papers exploring the crossmodal interactions between hearing and taste looked at ninety-four different studies that had been written up in sixty different articles. None of them deviated in any meaningful sense from what we've learned here. If you're still not convinced, and would like to read more of those sixty different articles, check out page 211 and knock yourself out.

Why do these correspondences exist?

There's a rather wonderful train of thought in science that is at once very serious, and also a great basis for a parlour game. If we follow Charles Darwin, any significant trait in any species of plant or animal should only be there if it confers some kind of evolutionary advantage to that species. This idea crops up in my research all the time. Apples, for example, are brightly coloured, shiny and seductive because that attracts animals, who eat the apple, carry on their way, and disperse apple seeds across a much wider area than the apple tree can reach on its own, ensuring those seeds don't grow into baby apple trees in the shadow of their parent, and compete for sunlight and nutrients.

I discovered my favourite example while researching my book *Miracle Brew*. I met a scientist at the Carlsberg Laboratory in Copenhagen who was researching why yeast sometimes released the butterscotch-like aroma compound known as diacetyl. What was it for? He eventually surmised that yeast does this when it's running out of food, and that fruit flies go crazy for diacetyl. So, when the yeast is in danger of starving to death, it sends up diacetyl like a distress beacon, the fruit flies come flying to get all the butterscotch yumminess from your beer, and then fly off again with yeast cells clinging onto their little hairy legs, hopefully to a new source of fermentable sugar.

Evolution is obviously a much slower process than societal or political change, which is why humans make so many dumb decisions that are unlikely to prolong our existence as a species. But the way our senses work would seem to be more hardwired. Is there an evolutionary advantage to crossmodal perception? There are various theories. All need more research.

For much of our existence, food supplies have been scarce. We need the energy that sugar brings, and we're highly tuned to notice it. Eating sweet, sugary foods gives us more pleasure than pretty much anything else, to the point where now we have too much access to it, we can't stop eating it when we should. In all the experiments conducted in crossmodal research on taste, sweetness is the most liked taste, and the strongest correlation between tones, timbre and musical style that we find pleasant.

This opens up a couple of possible explanations, and they aren't mutually exclusive. One is that crossmodal correspondences help the brain make decisions about what to eat more quickly than if we had to rely on rational thought. This could be instinct: if it looks like a duck, swims like a duck, and quacks like a duck, you could be reaching for the five-spice powder and Chinese pancakes before anyone else if crossmodal perception helps you figure that it really is a tasty duck quicker than other predators that like to eat duck.

Another suggestion is that we simply match things together that give us the most pleasure. If we like the taste of sweet things, and we like soft, mellow harmonies, we simply put them together. Or not-so-simply, in academia-speak, we indulge in a bit of 'hedonic transference'.

Or, it could be pattern-based. We love to make patterns in our attempts to understand the world. We like things to fit. So grouping anything that feels slow, heavy, dark, big, and low, and creating a separate group of anything that feels fast, light, bright, small, and high, through to thinking zingy and refreshing wine tastes more zingy and refreshing with zingy and refreshing music, could be patterns that we start learning as soon as we become aware of the outside world and start trying to make sense of it, whether that's about language, or pure sensory perception.

Whatever the main reasons are – for there are surely more than one – what's undeniable is that these effects are not random, and they are not rare. There's logic to how our senses overlap and interfere with each other, and that logic is widespread – maybe not globally, but certainly within any given culture. The potential this gives us for improving our enjoyment of beer by the judicious application of great music has hardly been scratched.

Bringing beer and music together – why do guitars (sometimes) taste like hops?

Having got to this point, we can see that beer and music play very similar roles in our lives and are linked in many different ways, even before we get to the spooky crossmodal stuff.

On a base level, beer and music belong together. A live gig without a beer is as wrong as a dive bar without a rock and roll soundtrack. But there's so much more than that.

Bands and musicians often use flavour words to describe their music. And when I'm writing about beer, I frequently use analogies from music. I often compare the four core ingredients of beer to the balance of guitar, drums, bass and vocals (if hops are lead guitar, a great axe solo always sounds better with a driving rhythm section behind it – which is why a hoppy beer should have a malty backbone). We talk about 'rock star brewers', which now has a dual meaning. At first, it meant brewers who gained a cult of personality within beer fandom that resembled that we'd normally think of in rock stars. But a few years after that, we had the flipside: bands and rock stars making beers of their own. Of all the things they could have done, suddenly, being photographed putting hops in the kettle was something any credible band had to do. Gary Barlow and Kylie Minogue may have their own wines, but Metallica, Motörhead, Elbow, Flaming Lips, Professor Green, Rush, Frank Turner and Mogwai are just some of the rock acts who have put their names to beers and got involved, to greater or lesser degrees, with the recipe and the brewing process.

Probably the most successful of these is Trooper, the beer created by Robinson's in collaboration with Iron Maiden. Robinson's marketing director asked me if I thought they should do it. I said no – the band thing was kind of over by that point. He didn't listen to me, and now it's Robinson's best-selling beer, loved all over the world. I've interviewed Bruce Dickinson, Maiden's lead singer and the brains behind the beer, several times. He is genuinely knowledgeable and passionate about beer, and takes the lead in formulating the recipes for what is now a wide range of Trooper beers.

Pubs and music are inextricably linked, too. There has to be a reason why our relatively small country punches so dramatically above its weight when it comes to producing globally influential rock and pop bands. I think that reason is the pub. It's not easy doing pubs gigs – I know, I've done a few myself. Engaging a half-drunk audience, a good chunk of whom wish you weren't there and think they could do better themselves, is both an anvil on which to hammer out your skills, and a crucible in which you can forge a devoted following. Without the pub, we simply wouldn't have the music scene we do. London venues such as Camden's Dublin Castle and Islington's Hope & Anchor are legendary in the history of late-twentieth-century music. As a sign of their 'authenticity', acts such as the Libertines, U2, Tom Jones, Ed Sheeran and Chris Martin have in the past few years gone back to their roots and played intimate pub gigs.

The endless debate around the definition of craft beer is identical in every respect to the prior argument over the true meaning of 'independent music', and before that, even punk. I've been filled in on this by my music writer friends, some of whom became friends after they saw an allure in craft beer that we had all seen in the 'alternative' music of the last quarter of the twentieth century. I met Andrew Harrison, who wrote about music for around thirty years and edited *Select, Q* and *Word* magazines, in 2010. At that time, he felt that the craft beer boom was 'Year Zero' for beer in the way 1978 had been for music. The rulebook had been torn up, and anything was possible. He was disillusioned with the music world by that point. As craft beer progressed from Year Zero, it seemed to me that Andrew's prediction became truer with every year that passed. Starting with his Year Zero analogy, and grossly oversimplifying everything that came after for cheap laughs, I plotted an uncannily similar trajectory.

	Post-Punk Music	UK Craft Beer
Year Zero	The Sex Pistols have destroyed the status quo. Anyone can pick up a guitar, learn three chords and form a band.	BrewDog have destroyed the status quo. Anyone can pick up a small brew kit, get their graphic design student mate to do some labels, and launch a craft beer brand.
The emergence of an anti-corporate, amateur aesthetic	The rise of indie labels. Sticking it to the man. Making amateurism, three chords, and not being able to sing into a virtue. Arguing over the definition of 'indie'.	The rise of craft beer brands. Sticking it to the multinationals. 'We're not quite happy with the recipe yet and it shouldn't have all that sediment. That'll be £6 please.' Arguing over the definition of 'craft beer'.
… Leading to an astonishing burst of creativity	Flowering of musical styles, incorporating a huge range of influences and turning them into something new. Bands like Talking Heads and Television combining musical skill with real innovation.	Flowering of beer styles, incorporating a huge range of influences and turning them into something new. Breweries like Kernel. Beer styles like barrel-aged imperial stout and Berlinerweisse.
Recognition of previous influences	'You know what? The first real punk record was actually "Louis Louis" by the Kingsmen, back in 1964.'	'You know what? The first recorded use of the term "craft beer" was actually in 1982, when Michael Jackson used it to describe Timothy Taylor's Landlord.'
Corporate bastards cashing in	Sigue Sigue Sputnik.	Aldi's 'Anti-establishment IPA'.

	Post-Punk Music	UK Craft Beer
Absorption by the mainstream	Sting's entire solo career. Especially 'Russians'.	Craft cans on heavy discount in Tesco, next to Stella and Fosters.
Oh, fuck it then	John Lydon advertises Country Life butter.	BrewDog co-founder James Watt attends Nigel Farage's sixtieth birthday party.
The final stage of grief: acceptance	'You know what? There are still some good tunes to have come out of it. Without punk we probably wouldn't have had Nirvana.'	'You know what? There are still some good beers to have come out of it. Neck Oil may be a pale shadow of what it was, but it's in every pub, and there was nothing like that so readily available before.'
'Sorry, who are you?'	A new generation arrives who don't remember what is was like before. They wear Joy Division and Sonic Youth T-shirts without even realising they were bands. Oh, wow, I just thought it was a cool T-shirt, yeah?'	A new generation arrives who don't remember what is was like before. They think hazy IPA is the only craft beer style, and send beers back to the bar if they're clear. They tell you IPA was invented in 2010 and has always been hazy, and if you didn't even know that, you clearly know nothing about craft beer.
Old man shouts at cloud	You realise that somehow you just turned into *that* guy, ranting about Joy Division novelty T-shirts and how the plural of vinyl is vinyl, and young people today don't know anything.	Old man shouts at cloudy beer.

Of course, there's more to music than post-punk and indie, and more to beer than craft. In their broadest sense, the similarities between beer and music run through them like two sticks of rock that have somehow fused together.

In both, individual tastes vary wildly – but overall, there's a rough, broad, critical consensus of what is good and what is bad.

Both are eternal – enjoyed throughout the world, and at least as old as civilisation.

We use both to change and heighten our emotional states, and as building blocks of culture and subculture, means of self-expression that signify of our personalities and identities, as individuals and as groups, and as something bigger.

Both are celebratory and euphoric, bringing us together communally. They both arouse huge passions in their fans. Both are always there for us at key moments in our lives: parties, anniversaries, important sports games, even funerals and wakes.

Beer and music are joined at the hip. So, it's really no surprise that the science is now proving that our enjoyment of beer and music don't work in isolation – the way we consume one can profoundly affect our impressions of the other.

That's it for the theory, the science and the explanation. Now it's time to turn to Side Two.

Crank up the volume, and crack open a beer. You've earned it.

SIDE TWO

· · · · · · · • · · · · · ·

The Practice

How Dizzee Rascal can help you overcome your dislike of sour beers.

And why U2 make it OK to secretly like Heineken.

The Principles of Pairing

SCAN HERE
FOR THE FULL
PLAYLIST

As we've seen, there are many different ways in which beer and music can be paired together. I've played around with different variations of the pairings listed here. Some of them overlap, and might be one or the other, while others can be more than one at once. The brain is a complex organ! And remember, it's meant to be fun.

The different principles of pairing don't all point in the same direction. Some of the science, plus years of doing events, suggest that there may be a particular association that's universal. But you as an individual might have a stronger, personal association that overrides it. For example, I might play a song that everyone in the audience loves – a stone-cold classic, with a beer they haven't tried before. Their feelings about the song will probably transfer over to the beer. That's called 'hedonic transference': the pleasure from one informs the other. But if it's a song you once got dumped to, you might not like the beer because of your own personal history, and that might override the associations pulling the other way.

Here's a list of different methods of pairing I've used here on Side Two, drawn from all the research on Side One:

- **Intuition and Feeling** The kind of pairing any of us might do when put on the spot. These two just feel right together. I started pairing tunes and beers like this before I knew anything about the science, and some of those early pairings still work brilliantly today. Although there may be more going on under the surface than we realise...

- **Context** Still something that's pretty simple and obvious. A beer and a song that you first encountered on holiday, or at a memorable party. A combination that seems to go together well with a particular mood, emotion or occasion.

- **Priming Words** I'm still not sure how this works. All I know is that it does – in proper scientific experiments, and in my own work. There are a lot of words that we use to describe more than one sense, or words that we use to describe sensations that are more associated with another sense. If I say both the beer and the song are heavy and dense, sunny and cheerful, minimalist, multi-layered – they're probably going to go together. Context is probably part of it. Maybe the power of suggestion is. But it wouldn't work if there wasn't something underlying all that in the way we parse sensory information.

- **Crossmodal Pitch Correspondences** The earliest work exploring possible correlations between flavour and music looked at whether different flavours corresponded with different sonic pitch. They totally do. It's obvious, even if you've never thought of it, which you probably haven't.

- **Crossmodal Taste Correspondences** Certain styles of music have been shown conclusively to map onto the basic tastes of sweet, sour, salty and bitter. This is fascinating in itself. What's even more mind-blowing is that changing the style of the music can alter your perception of the balance of these tastes.

- **Multiple Pairings** Both good music, and the flavour of beer, are complex and multi-layered. Sometimes two or three of these pairing principles can be working at the same time.

I'm not going to argue that these are the forty-five best beers in the world, or the forty-five best pieces of music, though I like them all. The main thing here is the *pairings*. I've chosen them to provide as broad an examination of beer and music matching, in terms of styles, genres and principles of pairing, as I can.

The beers

These are all beers I would happily drink – yes, even Heineken, if the mood and the context was right. (I've been a couple of times to Carnivale Brettanomyces, a festival of spontaneously fermented beers in Amsterdam. It's great. You should go. But after a whole day of drinking experimental wild-brewed concoctions, a cold pint of lager by the canal is heaven.) Each one is notable in some way. Most of them are readily available in the UK, although not all the time.

The playlist

The absolute joy of doing this book, and the events that led to it, was the obligation to explore music outside my normal comfort zone. Every song here is special in some way. They're all recognised as 'good'. A few are quite obscure; some very personal. Others, I discovered – or remembered – recently as good examples of what I needed for a pairing.

I've spread my musical net as wide as I can, but have stuck to the principle that, like the beers, they're all tunes I would happily listen to. There's no Queen, Led Zeppelin, Abba, Jay-Z, Coldplay or Ed Sheeran here because I don't like those artists, even though I recognise and accept that they are all talented and important members of the popular music canon. It's my book, and I can do what I want (to a point). If it makes you feel any better, my initial playlist contained over a hundred songs. I've had to exclude the Beatles, Doves, Elbow – *Elbow, FFS!* – Brian Eno, Big Country, Simple Minds, The Fall, and – *I can't believe I'm actually typing this, I am WEEPING right now* – Explosions in the Sky, because there simply weren't the right pairings available for them to go in the book. I only allowed myself one exception to this rule, and it's on page 184.

If you want to create your own playlist of music that isn't featured here, I've hopefully given you the tools to do so. Experiment – it's fun! The underlying principles really do work.

You can listen to those that made the cut, ordered not alphabetically, or by genre, or by method of pairing or anything like that, but in an order that just sounds right. This section follows the order of the playlist, track by track. I always used to make mix-tapes, and then playlists, for people I like. And I like you. This is the best one I've ever made, because there's a different beer to go with every track, and every one of those beers is wonderful. Please listen responsibly.

THE PAIRINGS

Duration Good Times with Sugababes 'I Bet You Look Good on the Dance Floor'

THE BEER

Duration, Good Times (4.2%), *American Light Lager*, UK

This beer exists because of a paradox in craft brewing. Craft grew out of homebrewing in the United States, with people creating beers in their garages that they couldn't find in shops or bars. Pretty much the only beers they could find in shops and bars were 'American light lagers', unchallenging brews which used adjuncts such as corn or maize instead of pure malt because they were cheaper, and because they created a lighter taste. The big brewers who dominated the market believed people didn't want their beer to taste of anything: they just wanted it to be cold, fizzy, and mildly intoxicating.

Craft beer proved them wrong. But craft beer is defined by innovation and experimentation: 'What haven't we done yet?' becomes a defining clarion call. Which means that if you have enough craft brewers, and enough time, eventually, like monkeys typing Shakespeare, some of them are going to realise that they haven't done American light lager yet. And so, they're compelled to recreate the very thing they once stood against.

Duration are one of my favourite breweries. Here, they've brewed a light, simple beer with corn as an adjunct to malt. They don't want you to overthink it. They just want you to drink it. Probably straight from the tin, as if you're some bro watching a ball game. Let's just say it's a lot better than the beers it was inspired by, slipping down easily and suggesting another one soon after.

Sugababes, 'I Bet You Look Good on the Dance Floor', Universal, 2006, *Pop*

The arrival of Arctic Monkeys marked a watershed in the evolution of the indie music I grew up with and defined myself by. This, their debut single, was one of the first songs to become famous via the internet, benefiting from the arrival of Apple's iTunes in 2004, and going on to become number one in the UK singles charts. Suddenly, everyone from the Chancellor of the Exchequer to the head of beer trade body Cask Marque was saying the band was 'rather good'. The idea of an underground cult or counter-culture seemed doomed. Your music taste couldn't be cool or obscure any more.

This seemed to be confirmed when Sugababes chose to cover the song as the B-side of their 2006 single 'Red Dress'. Couldn't miserable-looking teenage boys with floppy fringes and baggy sportswear have anything to themselves anymore?

So imagine my surprise when the cover turned out to be even better than the original. It's rockier. It's poppier. It's dancier. The girls obviously love the song and sing it with absolute sass. In the words of the TV talent shows of that era, 'they made it their own'.

THE PAIRING

Neither the beer nor the song is what you might expect. An 'American light lager' from one of the best craft breweries in the UK? An absolutely banging version of one of *the* indie anthems of the 21st century from a manufactured girl band? Both work *way* better than you would expect them to.

But this pairing is about more than that happenstance. I use it at the start of my live show, just before I go onstage. In my head, it's about a moment. That moment is Saturday night, 7.30. You've gone round to your mate's house to get ready before you hit the town. Make-up. Comparing clothes. Maybe a bit of pre-loading. It's decades since I've done this myself. But I imagine the thousands of women who have been in the Sugababes line-up at some point over the years doing it every weekend, leaving crushed cans of Good Times behind them when they're finally ready to go.

Hopback Summer Lightning with Primal Scream 'Slip Inside This House'

THE BEER

Hopback, Summer Lightning (5%), *Golden/Summer Ale*, UK

Back in the early noughties, people looking to make cask ale more relevant to a younger audience came up with the idea of 'golden ale': something that looked like a lager, was easy to drink, worked if served cold, but still had something of the character of a traditional ale. You can trace early golden ales back before this, but this is when they exploded: between 2001 and 2010, eight of the ten winners of CAMRA's Champion Beer of Britain could loosely be included in this style.

In 2001, the runner-up (to Oakham's JHB) was Hopback Summer Lightning. It was no stranger to CAMRA awards by this time, having been brewed since 1991 and winning CAMRA's Best Strong Beer award in its first year in competition.

Hopback describe it as a summer ale, and it's easy to see why. It's golden and clear, bright and fresh. The hops are zingy, the body light and fresh, belying the 5% ABV. It's a simple, straightforward beer in one sense. But something – probably caused by that relatively high ABV – keeps you coming back to detect hidden depths.

THE SONG

Primal Scream, 'Slip Inside This House', Creation, 1991, *Alternative Rock*

Primal Scream's album *Screamedelica* changed everything. Twee indie evolved into a melange of rock, acid house, psychedelia and dub. Past, present and future seemed to coalesce into a technicolour melting pot. It was, by any measure, the standout album of 1991.

'Slip Inside This House', the album's second track, is a cover of a 1967 song by psychedelic rock band the 13th Floor Elevators. The original is a fuzzy, grungy groove with lyrics about enlightenment via a mish-mash of religious and spiritual beliefs.

On *Screamedelica*, Andrew Weatherall's production opens it up. As anyone with the possible exception of Primal Scream's Bobby Gillespie will admit, this is really Weatherall's album rather than theirs, a reminder of the time when DJs and producers were just emerging as stars in their own right. If 13th Floor Elevators were trying to get you into a church, with Weatherall you're already in it. But now it's home to a party rather than a religious service, or maybe a party *as* religious service. That party is well under way. It's echoey and groovy and shambling and euphoric, and fills your mind with light. No, I haven't been taking drugs. But the song makes you feel like you have.

THE PAIRING

There's a day each year which is the real first day of spring. Now the weather is so completely random, it could fall any time between early March and mid-May. Whenever it falls, it's the first sunny day that's not icy cold, the first temperate day that's not shrouded in cloud. Your body responds to it by waking up. You turn your face to the sun, almost like a sunflower. It's the first day you can drink in a pub beer garden – or indeed your own garden – without a coat. If you're at home, it's the time to throw open doors and windows, and leave them open without anyone questioning your sanity. When this day arrives, I always reach for the blissed-out cartoon sun on the sleeve of *Screamedelica*. 'Slip Inside This House' is an invitation to the sun and the warmth. The world is opening back up after hibernation. And Summer Lightning is a taste of things to come. Despite the name, that extra strength and hidden depth to the beer makes me think of it not as a midsummer, baking-hot August beer. It's a beer that *announces the arrival* of summer, with some of the cool, loamy, dew-moistened spring still knocking about in there somewhere.

Hoegaarden with Neil Young 'Harvest Moon'

THE BEER

Hoegaarden (4.9%), *Wheat Beer*, Belgium

The beer that introduced Britain to Belgian wit, when it arrived in the UK in the late 1990s, Hoegaarden was a preview of the craft beer boom that would hit a decade later. It was different. It was … *tasty*. It made people re-evaluate what beer could be.

Hoegaarden may now be owned by AB-InBev, the pantomime villains of the beer world, and they may have tried to ruin it, like they ruin every beer they touch, but thanks to pressure from Belgian beer fans, it's still a cracking beer and a great example of a Belgian witbier. Thanks to the grist being about thirty per cent wheat, combining with a particular yeast that is partial to such a wheat-heavy mix, Hoegaarden is pale yellow and naturally hazy, with a smooth, almost creamy body. The addition of coriander seeds and orange peel lends an engaging spiciness to the slightly sweet and sour flavour. A perfect summer beer and a versatile food pairing. I tend not to choose wheat beer myself when I'm drinking. But every time I use this for a tasting event, and have a little sip, I wonder why I don't drink it much more often.

THE SONG

Neil Young, 'Harvest Moon', Reprise, 1992, *Folk Rock/Country Rock*

After the heavy *Ragged Glory* album left Young with tinnitus, he returned to his acoustic roots for this beautiful, country-tinged album, a sort-of-sequel to 1972's *Harvest*. He was forty-seven years old at the time. When I came across it, that seemed unimaginably ancient.

Lyrically, the album is about love and relationships, with a reflective stance born of maturity.

The title track is a sentimental slow dance in a dark corner, a dewy-eyed celebration of enduring love. Perhaps he's singing to his wife of many years, as the official music video suggests. Or maybe he's watching an old flame from across the room, thinking of their time together and saying to her that it's not too late to reignite their love, which is what the lyrics

suggest to me. Either way, it sounds like it's being sung about five minutes after it was written, around a campfire in a field or a barn softly illuminated by fairy lights and fireflies, on one of those magically warm, late summer nights when anything could happen.

THE PAIRING

When I first started thinking about pairing beer and music, I posted a simple tweet, with no context or explanation: 'What kind of music goes with wheat beer?' The answers were plentiful and remarkably consistent: alt-country, bluegrass, folk, and specifically, Bert Jansch or Neil Young. People sensed a relationship between wheat beer and country-ish music, even if they hadn't thought about it until I asked the question.

I'd always wanted to pair something with this song, so it seemed like an obvious choice. Since then, it's always been the first pairing I try on people in my events, to get them used to the whole idea. It's about context, and that's an easy route into the whole idea of beer and music matching. But how exactly does it work? Does the word 'wheat' evoke a sense of countryside vistas? Does the golden haze of the beer remind people of the summer sun? Or does the smooth, languid rhythm of the track suggest the kind of lazy day when you want to kick back and enjoy a refreshing, smooth beer like this? Both the song and the beer evoke long, hazy summer evenings, the sun lowering in the sky while a blissed-out party happens among the hay bales after a long day doing nothing.

Asahi Super Dry with Kraftwerk 'Trans-Europe Express'

THE BEER

Asahi, Super Dry (5%), *Lager*, Japan

I was once lucky enough to be given a personal guided tour of Asahi's main brewery, just outside Tokyo. My lasting memory of it is polished steel. It looked like a brewery imagined by a science fiction writer. Everything was gliding and smooth and futuristic. At one point, a shining metal cylinder the height of the room began a slow descent, until the top of it was waist height. Inside, a circular screen then showed film of beer yeast fermenting for a minute or so. Then, with a low hum, the cylinder rose again, carrying its *saccharomyces cerevisiae* back to the ceiling.

The beer itself feels the same way, both in its character and its branding. Western designers always look to Japan for new trends, and I remember when Asahi first appeared in the UK, a while after its initial 1987 launch. It was a beer you just had to be seen with: silvery with Japanese script, it was as futuristic as the brewery it came from.

I'm sure a seasoned ale drinker would dismiss Asahi as tasteless fizz: the point of a 'super dry' lager – a concept invented with this beer – is that the sugars are more highly attenuated than even other lagers, leaving a clean, crisp, dry taste with a short finish.

In a broad palate of beers, this light, dry character does have its place. When we think about Japanese cuisine, we probably think of sushi or ramen first. But in the *izakaya* (small cafés or bars) beneath Tokyo's railway arches, a lot of the snacks that people eat with beer are barbecued in rich, sweet glazes, or deep-fried. Asahi Super Dry cuts through the grease perfectly. Don't think of it as tasteless. Think of it as artfully minimalist.

Kraftwerk, 'Trans-Europe Express', Parlophone, 1977, *Electronic*

Kraftwerk have been called the most influential band since the Beatles, and this is often justified with reference to the 1977 album *Trans-Europe Express*, and in particular to the title track. The mood of the album is of a joyously optimistic future – or seems to be. Here and there, the music hints at something darker behind the polished surfaces.

The track itself is usually described as a celebration of travel across a continent which, only thirty years earlier, had been convulsed by world war. The ease with which the long train navigates the continent opens up endless possibilities: the sparse lyrics mention breakfast on the Champs Elysée followed by tea in Vienna, before nipping over to Düsseldorf to hang out with David Bowie and Iggy Pop.

Musically, there's nothing here that isn't needed to create the sense of travel. Minimal electronic beats and a simple, reedy keyboard riff create a sense of space that lives between the notes, not in them.

But the Doppler effect as the train goes past suggests we might not be on it. The visuals associated with the track show a seemingly endless train bearing down on us as we watch from the trackside. And there's a hint of something sinister in the minor key melody. There's menace here as well as possibility.

THE PAIRING

This is one of the first two or three pairings I thought of when I had the idea to match beer and music. Both the song and the beer are about straight lines and efficiency. The album cover and the beer packaging could almost have come from the same designer. Created ten years apart, they share a slightly dated futurism that no longer exists – imagining a future that's more optimistic than we foresee now, but perhaps sacrificing a bit of human warmth and wonkiness in favour of machine-controlled efficiency.

Young's Double Chocolate Stout with John Grant 'GMF'

THE BEER

Young's, Double Chocolate Stout (5.2%), *Chocolate Stout,* UK

While I love the kaleidoscope of aromas that hops can bring to a beer, I'm always looking for opportunities to press the case that malted barley contributes just as much, albeit in not quite as showy a way.

What I love about the flavours of malt are that they're a product of human ingenuity. You take the same barley grain, and depending on what you do to it, it can taste of digestive biscuits, toffee, caramel, red berry fruit, coffee – or chocolate.

Back when Young's was still a proud brewery in Wandsworth rather than a cluster of dying brands owned by a company that has absolutely no interest in nurturing them, they brewed a variety of interesting beers. This – at the time of writing at least – is the last survivor of that playful tradition.

You like the chocolate notes provided by highly kilned chocolate malt? Cool! Well what if we enhance those flavours with ... more, actual real chocolate! Some might call it overkill. Others call the combination of chocolate malt, real dark chocolate and chocolate essence (hang on, shouldn't it be Triple Chocolate Stout?) decadent, indulgent, luxurious, or just bloody lovely. Importantly, and miraculously, all that chocolate does not turn this into some ghastly, over-sweet chocolate liqueur-type travesty of a beer. It's smooth and drinkable, and doesn't get cloying. I just hope it still exists by the time you read this.

John Grant, 'GMF', Bella Union, 2013, *Pop*

I first got into John Grant when he was the lead singer and songwriter for a band called The Czars. Their songs were bittersweet and sad, shot through with wry, reflective humour. Like many others before them, they did a cover version of

Tim Buckley's 'Song to the Siren', possibly my favourite song of all time. The arrangement, but more than that, Grant's pure, plaintive voice, made it one of the best versions I've ever heard. So I was sad when the band split in 2006.

Four years later, Grant was back as a solo artist, his voice now more baritone than tenor, his lyrics more intensely personal, his music featuring a bigger synth element. 'GMF' is from the second solo album, 2013's *Pale Green Ghosts*. The lyrics are a form of courtship, wrapped up in an ironic self-appraisal, the verses reflective and sometimes self-critical between the chorus where Grant boasts that he is the 'Greatest Motherfucker that you're ever gonna meet.' It's funny, sweet, touching, and you'll find yourself singing it out loud in places you probably shouldn't.

THE PAIRING

The research into crossmodal correspondences between different senses began by looking at synaesthesia, the condition by which some people experience sensory information via the 'wrong' channels – seeing music, or smelling words. It soon became apparent that what we're talking about in this book is different from synaesthesia. But part of my reaction to this song feels close to it: *John Grant's voice sounds like rich, gooey, melted chocolate*. It's a simple as that. I hope you hear it too.

Even if you don't, you would hopefully agree that certain words can be applied to both the voice and the chocolatey character of the beer. Smooth. Silky. Rich. In terms of taste mapping, there are some 'sweet' elements here: the length of the sung phrases, and the relative quietness of the song map on to sweetness. But at the same time, the low pitch brings forward the gentle bitterness that's also in the beer.

But yeah, whatever. That voice. Chocolate. That's it. That's the pairing.

Weihenstephan Hefe Weissbier with Driven Snow 'When You Sleep'

THE BEER

Weihenstephan, Hefe Weiss (5.4%), *Wheat Beer*, Germany

Weihenstephan claims to be the world's oldest brewery. When anyone says this, it's almost certainly not true, or is at best unverifiable. But why does it matter? It's been here for centuries, and learned over that time how to make truly excellent wheat beer. Don't just take my word for it – pretty much every time I'm involved in judging beers awards, this one medals.

German wheat or white (*Weiss*) beer is a close relative to the Belgian version, but presents quite differently. Coming as it does from the home of the German *Reinheitsgebot* purity law, it eschews added fruit and spices. (Wheat was added to the text of the law some years after it was first issued, after pressure from the Bavarian princes who enjoyed a wheat beer.) But it showcases the expertise of the German brewing tradition by getting fruity, spicy flavours into the beer from its core ingredients. It's years since I've had banoffee pie. It feels like a very late-eighties dessert. But I'm reminded of it every time I taste this beer. Banana and clove aromas come from the yeast, with the malt base adding the crust of the banoffee pie.

As a style, it's an extraordinarily versatile food pairer, going with anything citrusy or spicy. It's similarly versatile as a match for music, with elements of sweetness, sharpness, a hint of bitterness, and an interesting, smooth mouthfeel and texture.

THE SONG

Driven Snow, 'When You Sleep', Driven Snow Music, 2024, *Indie Pop*

Spotify has its faults, and they are major. But having unwittingly trained its algorithm in the kinds of music I like, by making regular playlists and allowing it to suggest more stuff that would fit, it occasionally throws up some absolute treasures.

'When You Sleep' was a track from My Bloody Valentine's revolutionary 1991 album, *Loveless*. Even today, the album is a thrilling listen. The constant is feedback-drenched noise. The variation is the sheer number of different directions that noise gets taken in, the variety of different sounds you'd never heard before. The hook in 'When You Sleep' is one of the strongest on the record, a dying robot trying to sing an alien distress signal.

Thirty years later, Driven Snow disinterred the melody from under the rubble for their debut album. Kieran McGuinness and Emily Aylmer had each been in bands around Dublin that had been well thought of but hadn't cut through. They settled down to have children, and started writing soft, gentle music during lockdown, after the kids were in bed. They realised they had something worth sharing with the world, and released an album, *A Kind of Dreaming*, early in 2024.

'When You Sleep' is the only cover version on it. The melody is picked out delicately as a simple piano riff. The lyrics are revealed as an achingly sweet pop song. Guitar, drums and bass provide a soft, latent power that stops this from being one of those fey songs that were always used in ads for mobile phone networks in the 2000s. It's exquisite.

THE PAIRING

This is an example of how sweetness in beer pairs with melodic, high-pitched music. It's there in the piano line, and it's there in the beautiful vocals that don't exactly harmonise, but weave around each other, wrapping themselves together.

It's there in the beer, in the lemon and banana hints. These, for me, are accentuated when the music is playing. It's not that the sweetness becomes more intense, but it does become more noticeable.

The beer never becomes too sweet though. This is the trick of skilful brewing. The sweet elements are anchored by everything else. That's echoed by a song that could be very saccharine in the wrong hands. The treatment of it here allows those 'sweet' elements to shine, but they're countered by a really strong rhythm section that suggests these guys could rock if they wanted to, but this is not the time or place for it.

Magic Rock Salty Kiss with Television 'Marquee Moon'

THE BEER

Magic Rock, Salty Kiss (4.1%), *Gose*, UK

Magic Rock was founded in 2011 by Richard Burhouse and head brewer Stuart Ross. When I first met Stu, he was brewing on a homemade kit in the basement of the Hillsborough Hotel in Sheffield. It was thrilling to see what he did when he was let loose on a proper brewhouse. Magic Rock quickly grew to be one of the darlings of the first wave of British craft beer.

In 2015, they did a collaboration with Danish brewer Anders Kissmeyer, which turned into Salty Kiss. It's a German Gose, a style that was almost extinct until the early 2010s. Brewed specifically in the town of Goslar in Saxony, legend has it that after the Second World War only two living people knew how to brew it. Gose is brewed with salt. Kissmeyer brought sea buckthorn for a sour snap, and they added gooseberry for good measure.

Salty Kiss is tart and refreshing. Salt may sound like you're going to be drinking sea water, but instead, it brings a selzerish dryness that complements the sourness perfectly and creates a very drinkable beer.

This may all sound to a traditional ale drinker like its craft beer jumping the shark. On paper, it could be seen as weird for weird's sake, a classic example of the Jurassic Park line, 'Your scientists were so preoccupied with whether they could, they didn't stop to think if they should.' It's a line I use a lot in the craft beer world these days. But not with reference to Salty Kiss. You don't have to take my word for it. Since it's 2015 launch, it's been one of Magic Rock's best-selling beers, a solid member of their core range.

THE SONG

Television, 'Marquee Moon', Elektra/Asylum Records, 1977, *Alternative Rock*

Like his friend Patti Smith, Tom Verlaine originally moved to New York to be a poet. He borrowed his surname from French Symbolist poet Paul Verlaine, and adopted the poet's style of vivid, dreamlike imagery for his lyrics.

It's tempting to refer to his band Television as post-punk, but many of the songs on their debut album, including this, its title track, preceded punk. It was Verlaine who persuaded Hilly Kristal, the owner of legendary club CBGBs, to book acts outside their original remit of country, bluegrass and blues, and it was from CBGBs that New York's legendary punk scene emerged.

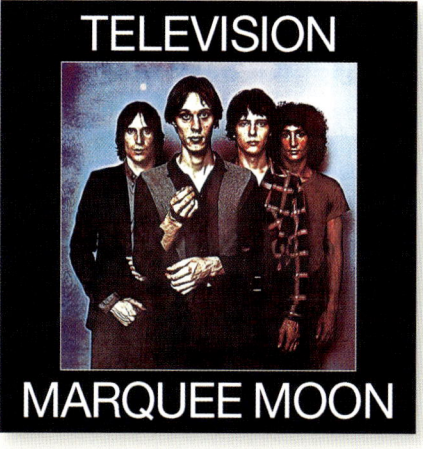

The lyrics of 'Marquee Moon' are beguiling and haunting, but it's the guitars that make it imortal. The whole song is an interplay between the virtuosity of Verlaine and fellow guitarist Richard Lloyd. It's fascinating because although the guitar solos demonstrate the two are very proficient players (which always made them feel apart from the punk scene) the basic driving rhythm consists of just two notes, but a very complex arrangement builds around them. The rising break that begins at 8'14", and just continues to build and build creating a twisted tension before a glorious release, remains genuinely thrilling no matter how many times you hear it.

THE PAIRING

This is a textbook pairing from the emerging field of neuro-gastronomy. The pioneering work done by Bruno Mesz and others, since backed up by further scientific work and my own semi-serious aping of them, suggests that salty flavours pair with music that has short notes and a staccato articulation. I can think of no better example of this than the choppy, angular, two-note melody in 'Marquee Moon'. I use the introduction to the song in the exercise where I ask people to match tastes with music styles. A third of respondents pair it as salty, a far higher correlation than if it were a random choice. Does the music make the beer taste better? I don't know – you'd have to ask someone who doesn't think it tastes glorious to begin with.

Badger Brewery Outland West Coast IPA with The Killers 'Human'

THE BEER

Badger Brewery, Outland West Coast IPA (5%), IPA, UK

There's a long-running debate in British beer that reveals an inherent shallowness of the craft beer scene: is it OK for an old-school, traditional brewer to release a modern craft beer style?

It's an absurd argument when you think about it. Were brown ale brewers accused of dad-dancing when they tried their hand at porter in the late eighteenth century? I suppose it's one thing if a Victorian brewery suddenly starts putting cartoon skeletons or lupuloid monsters on their packaging – that's not them. But I have absolutely no problem with a brewery founded in 1777 launching a West Coast America IPA. Especially when it tastes like this.

What we now call 'West Coast IPA' was, until about 2012, just 'IPA'. But ask for IPA now, and you're likely to get the dominant New England (NEIPA) style. At the time of writing, there is a bit of a revival of West Coast IPAs, but many of these seem to be created by people who have only ever known hazy, juicy NEIPAs, and are working back from there: make it a bit less hazy and juicy, and a little more bitter.

Badger's Outland starts with what a 'Westie' was twenty years ago. It's orange-brown, clear, has a lot of citrus fruit and pine resin on the nose, a toffee-ish malty backbone to balance all the hops, and a clean, bitter finish.

THE SONG

The Killers, 'Human', Island Records, 2004, Pop

It's odd that a rock band from Las Vegas helped define the British cultural fabric of the first decade of the twenty-first century. It's as if *Gavin and Stacey* had been written by Germans, or *Shaun of the Dead* was made in China.

But two songs by the Killers define that decade musically. The status of the first hit, in 2004, was immortalised fifteen years later in possibly the greatest ever headline from spoof new website *The Daily Mash*: 'Twelve dead after hen party hears first notes of "Mr Brightside"'.

There were other hits. But in 2008, 'Human' achieved almost the same unofficial national anthem status. When I was researching my 2022 book *Clubland*, I visited working men's clubs across the country. At every single one, where there was a karaoke night, people were fighting to sing 'Human'. What is perhaps more remarkable is that most of those doing so were in their late seventies and early eighties.

It's not that younger people don't like the song; it's more that it seems to possess some strange force that gets everyone from small children to people on walking frames onto the dance floor. It's infectious. It has a naggingly strange singalong lyric in the chorus.And it's somehow become as much a part of British culture as Wallace and Gromit having a pint of real ale from a dimpled jug in the Rover's Return.

THE PAIRING

There's a contextual link here about American imports and British culture, especially as in both cases it's not a one-way street: both IPA and the indie rock of The Killers are examples of the cultural transatlantic tennis match where we lob something their way, they change it and lob it back, and so on. The Killers grew up listening to New Order. American IPA began life as an attempt to recreate the British IPAs of the nineteenth century.

That all fits, but it's not the reason for the pairing. I've always loved bands that play off sweeping synths with crunchy guitars, and I like a beer that's a great balance between the solid backbone of the malt and the pyrotechnics of hops. Some craft beer fiends think of balance in beer as boring. But a perfect balance can be high or low, loud or quiet, big or small. It's just about everything working in harmony. And harmony is a good thing.

Guinness with Richard Hawley 'Open Up Your Door'

THE BEER

Guinness (4.1%), *Stout,* Ireland

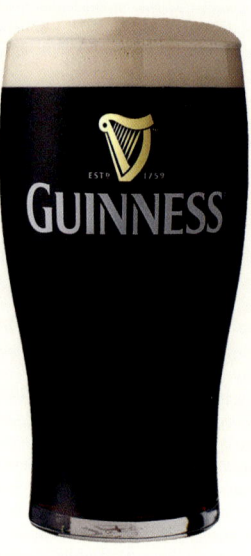

What can we say about Guinness that hasn't already been said? It's not even in the top ten of the world's most valuable beer brands, but it's easily the world's most iconic beer, in an age when we seem to be forgetting what 'iconic' really means.

It's a beer people aspire to drinking. The fact that Guinness is Ireland's national drink means that, for example, Japanese people drink it to 'gain permission' to behave a bit more like the Irish do. It's part of the reason – the Irish diaspora notwithstanding – why Ireland's national saint day is celebrated around the world.

The Nigerian version was for years marketed with a rather unsubtle suggestion that Guinness made you 'more of a man', that it 'put lead in your pencil'. This has been cleaned up and made a little more ambiguous in the long-running global 'Made of More' advertising campaign – if you drink Guinness, you just have a bit more substance about you.

That's because while people want to be seen as Guinness drinkers, they're afraid of the dark. People still refer to Guinness as 'a meal in a glass'. It pulls off the neat trick of looking like a 'grown-up' drink that beginners would struggle with. But when you take a sip, it lets you into a secret – it's really easy to drink. A velvet fist in an iron glove.

In the great Guinness drought of 2024–25, beer geeks were queuing up to tell anyone who would listen that there are far better porters and stouts out there. If by 'better' you mean more flavourful and interesting, this is true. But that missed the point of why many people drink Guinness. It's a triumph of branding more than brewing. A beer to be seen with. And I don't mean that as a criticism.

Richard Hawley, 'Open Up Your Door', Parlophone, 2009, *Alternative Pop*

Richard Hawley is from Sheffield, just down the road from where I grew up. Where he's from in time is less clear. He's roughly the same age as me, but seems born in some other era. His velvet croon, perfected, in his own words, with 'booze and fags', seems more at home any time between the 1940s and 1980s than it does now.

My teenage Sunday afternoons were spent in my bedroom doing homework. My dad invariably worked as much overtime as he could. As an early Hawley song title had it, 'I'm on nights'. He was on weekends too. In dad's absence, as the smells of roast beef and cabbage, ready for his return, wafted up to my room, mum would play Billy Fury, and other crooners from the late fifties and early sixties. My memory attempts to trick me that Hawley was among them, even though he too would have been just a kid at the time.

'Open up Your Door' stands out in an impressive field as one of his best. Certainly, it's archetypal Hawley, a perfect single-song intro to his work. Lovelorn and romantic, beseeching and timeless. If you were its subject, you'd be reaching for the handle after the first line.

THE PAIRING

Sheffield artist Pete McKee draws simple but evocative cartoons of local icons, both real people and archetypes. He's drawn Hawley several times. His most famous image shows the singer hunched at a pub table, guitar case behind him, dog at his feet, fag in his mouth and a pint of Guinness in his hand.

Would I have made this pairing if I didn't know that Hawley conducts pretty much every press interview over several pints of Guinness? I know several people who have been in these sessions with him, and am jealous of them all.

His personal connection to the beer makes it an obvious match. But there's more than that. Hawley's songs and characters inhabit the same backstreet boozers he does. And if you drank in them, you'd probably order a Guinness too. Both the song and the beer are redolent of mid-afternoon snugs, the sun slanting in catching dust motes, the smell of last night's spilled pints, wistfulness and longing still on the air.

Sierra Nevada Torpedo with Ronettes 'Be My Baby' and Billie Eilish 'No Time to Die'

THE BEER

Sierra Nevada, Torpedo Extra IPA (7.2%), *IPA, USA*

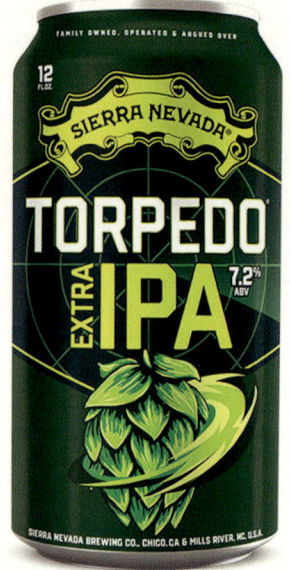

For many drinkers, Sierra Nevada Pale Ale sits at the centre of the craft beer map. It's one of the beers that set the course for the modern American craft beer revolution. From any extreme, you can come back to the centre and always find Pale Ale, as perfect as it has always been.

But craft beer moved on. By the late noughties, a beer that had once blown people's minds with its hoppy intensity now seemed a little ordinary. So, Sierra Nevada innovated again. In 2009, they invented the Hop Torpedo, a tall cylinder with which they took the established British cask ale practice of dry-hopping (adding fresh hops to the container holding a finished beer) to a new level. The Torpedo is hand-filled with 75lb of whole-cone hops. Then, beer from the fermentation tank is circulated through this dense column of hops before returning to the tank.

The base Torpedo beer is 7.2% ABV, with sweet maltiness balancing an intense hop bitterness. The Torpedo process massively intensifies and broadens out the aroma and flavour from the hops, with big swathes of tropical fruit, citrus and pine and a long, buzzing finish. In its way, it's as balanced as Pale Ale. But balance can be at the top of the scale as well as lower down.

The Ronettes, 'Be My Baby', Philles Records, 1963, *Pop/R&B*

The sweet second hit for the Ronettes has so many firsts, so many records, an entire book could be written about its story. It's been played on TV and radio millions of times, and influenced many of the greatest musicians who came after it. It was the first time that legendary producer Phil Spector used an orchestra on one of his records, and became the blueprint for his legendary 'wall of sound' production. The mighty, echoing 'Bum…da-Bum-BOOM' drum intro has been copied hundreds of times, and has become one of the definitive drum patterns of pop.

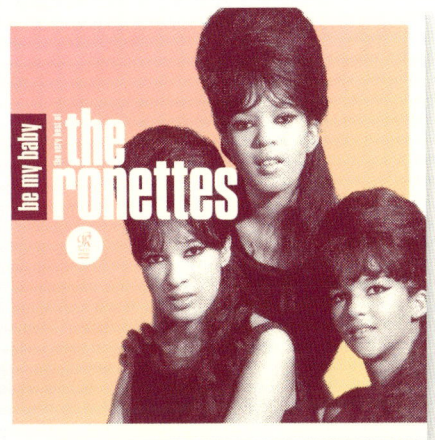

Sisters Veronica ('Ronnie') and Estelle Bennett, and their cousin Nedra Talley, began performing together as children at family events, already perfecting their trademark three-part harmonies before their teens. They formed the Ronettes in 1957, when Ronnie was fourteen, and had their first record deal in 1961. When they auditioned for Phil Spector in 1963, he leapt from his chair and cried, 'That's it! That's it! That's the voice I've been looking for!'

Spector, whose marriage was failing, fell in love with Ronnie – they would be married from the late 1960s to the early 1970s. Allegedly, he wrote the lyrics of 'Be My Baby' as a declaration of his feelings for her.

This can be read in two ways. Firstly, as a slightly sinister exercise of power by Spector, making the object of his desire sing his feelings back to him. But at the same time, this was the reversal of a dominant pop trope. Usually, it's the woman who is the 'baby' in songs, with a male singer infantilising her and seeking to control her. Whatever Spector's intentions, this role reversal, where a teenager woman with a high, girly voice is asking a man to 'be my little baby', was quietly revolutionary.

THE SONG (2)

Billie Eilish, 'No Time to Die', Darkroom/Interscope Records, 2020, *Orchestral Pop*

Like Ronnie Spector, Billie Eilish was still at school when she began creating music. She was just fourteen when she released her first single, written and produced by her brother, Finneas O'Connell. Her first studio album, *When We Fall Asleep, Where Do We Go?*, was released in 2019, making her a global star and one of the highest-selling artists of the 2010s. Records quickly began tumbling from then on.

Billie and Finneas collaborate closely on everything, and claim to have written several songs together that they imagined as themes to James Bond films. When Eilish was seventeen, she was given the opportunity to create one for real.

'No Time to Die' is the theme song from the Bond film of the same name. Eilish is the youngest person ever to have recorded a Bond theme. When it debuted at number one in the UK charts, she became the first artist born in the twenty-first century to have a number one hit. This record was echoed when Eilish became the first artist born in the twenty-first century to win an Oscar, with 'No Time to Die' being named Best Original Song in a movie.

Eilish's smoky, half-whispered, heartbroken vocal has a vulnerability that's at odds with the famous Shirley Bassey belters of previous Bonds, but in its way it's just as powerful. Hans Zimmer's stirring string crescendos and Johnny Marr's guitar twang create the trademark James Bond drama, and the whole thing comes together with a sense of foreboding that foreshadows the path of the film.

THE PAIRING

A few years ago, a pop-up restaurant called House of Wolf in Islington involved diners in an experiment that explored crossmodal pairing. They served a bittersweet lollipop: bitter coffee ice cream wrapped in a sweet chocolate coating dotted with sugary sprinkles. In front of the diners was an old-fashioned telephone: you could dial one, and get an audio accompaniment that supposedly emphasised the bitterness in the lollipop; two, to emphasise the sweetness.

That's what we're recreating here. Sierra Nevada Torpedo is a powerful beer that delivers both bitterness and sweetness in heavy doses, making it the perfect beer to recreate a real scientific experiment in crossmodal pairing. When hops are added at the start of the boil, they undergo a process called isomerization, creating bitterness. Add hops at the end or even after the boil, and their precious aroma compounds stay intact. The Hop Torpedo turns up the volume on hop aromas massively, giving the beer its sweet citrus and tropical fruity notes.

Using the same technique as House of Wolf, one of these song aims to accentuate the sweeter notes in the beer, while the other brings out the bitterness.

But which way around? Try for yourself before reading on.

Is there any difference in how you perceive the beer when the song changes? If so, which one is bitter and which is sweet?

According to the experiments carried out by Bruno Mesz detailed on Side One, the 'sound' of sweetness is high-pitched and melodic. One of the best examples of that is a close vocal harmony, and 'Be My Baby' fits the bill perfectly.

Meanwhile, bitterness is low and sonorous, *legato* in musical terms. The sound of a cello or double bass being played with a bow captures it perfectly for me.

I used to use Soul II Soul's 'Keep on Moving' here. It's deep and groovy. But it also contains quite a bit that pairs with sweetness, so it isn't great for this direct comparison. 'No Time to Die,' with Eilish's deep, breathy vocal, those bold, swirling strings and the dark, powerful mood of Bond themes in general, should prove the perfect accompaniment to the bitterness of the hops that builds at the finish.

In my experiments, it's not the same way around for everybody. But almost everyone perceives a difference, and a clear majority feel the Ronettes bring out the sweetness, Eilish the bitterness.

Robinsons Old Tom with Joy Division 'Transmission'

THE BEER

Robinsons, Old Tom (8.5%), *Barley Wine*, UK

First brewed in 1899, this barley wine was inspired by a sketch of the Robinsons brewery cat, by head brewer Fred Munton. He did the sketch at the top of the page on which he wrote the recipe. The fact that that recipe book is preserved today means we can see that the beer hasn't changed since.

Old Tom is a throwback to brewing history in more ways than one. It was brewed at a time when hops were used primarily for their bitterness rather than the fruity aromas we prize today. So this is a beer that's all about the malt. And with enough malt for the fermentable sugars to give 8.5% ABV, there's a symphony of chocolate, roasted nuts, smoke, molasses, liquorice, and vinous dark fruits. It's not just a sipper because of the strength: it's a beer that rewards time taken to savour.

It also reminds me of a time before the craft beer revolution, when British brewers created beers that were generally very similar – bitters and pale ales between 3.8% and 4.2% ABV. But every now and again – for Christmas, a royal wedding or birth, or a brewery anniversary – they'd wheel out big guns like this. They wouldn't always be easy to find, but they were worth seeking out. Incredibly, Old Tom was only available on draught until the 1930s. That fact that it's now constantly available in bottles is always a cheering thought.

THE SONG

Joy Division, 'Transmission', Factory Records, 1979, *Alternative Rock*

'Transmission' was Joy Division's first single. They'd been together for a while by then, playing live and abandoning a debut LP because it wasn't as good as they knew they could be. The gigs made them an incredibly tight band. The eventual first album, *Unknown Pleasures*, which was recorded after they'd done a first, slower tempo version of 'Transmission', defined their sound. After the album's release, they performed a reborn 'Transmission' on the Granada TV show *Something Else* in September 1979. It remains utterly compelling. There's never been a rhythm section to match Joy Division: the drums and base are not just percussion; they drive the song forward at a relentless pace while creating a brooding, menacing

atmosphere, which in turn is slashed by Bernard Sumner's jagged guitar lines. Ian Curtis's vocal builds from mere urgency to a sense that his life depends on him delivering the song. Incredibly, Curtis would die by suicide just eight months later.

THE PAIRING

This is the perfect pairing to demonstrate context in beer matching, and also how words matter and can 'prime' the listening drinker, or drinking listener.

Hearing about the Heriot-Watt University experiments about pairing 'zingy and refreshing' wines with 'zingy and refreshing' music, and 'powerful and heavy' wines with 'powerful and heavy' music, I started thinking about how to steal – sorry, apply – the same principle for beer. To an extent, both 'Transmission' and Old Tom could be described using the same words. Both are strong and powerful. Both are dark, intense, and brooding. Menacing, even.

Then, this contextual, linguistic relationship takes on a further layer of meaning. Because both come from the same town. Stockport, just outside Manchester, is a post-industrial town, home to both Robinsons brewery and 10cc's Strawberry Studios, where 'Transmission' and *Unknown Pleasures* were recorded.

By the seventeenth century, Stockport was a centre of the British hat-making industry, and later the silk industry. It industrialised early, and hard. In 1844, philosopher Friedrich Engels wrote that it was 'renowned as one of the duskiest, smokiest holes' in the whole of the industrial area. In places like this, the dirt and smoke always seem to outlive the prosperity that industry brings. Both the beer and the song feel like they reflect their post-industrial terroir. It's hard to imagine either coming from the golden fields of, say, Norfolk.

Verdant Light Bulb with Delays 'You Wear the Sun'

THE BEER

Verdant, Light Bulb (4.5%), *Hazy Pale*, UK

My problem with hazy, juicy pale ales is not the style itself. It's that the style is very easy to make, and mediocre brewers can just churn them out, and people will buy them. If those crappy, shiny, slightly unnerving AI pictures on social media were a beer style, they would be hazy, juicy pale ale.

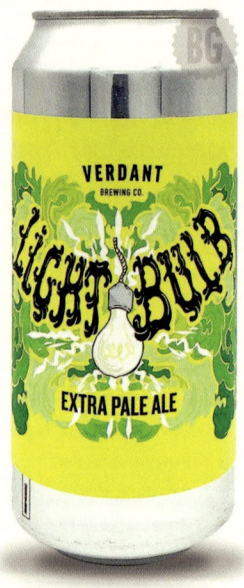

That analogy would make Verdant one of the human artists whose work is being stolen by AI. When they designed this beer, Verdant did not say, 'We need to brew one of these because everyone else is.' They wanted to create an extra pale ale that had the juiciness they personally liked, but also the drinkability that meant anyone could enjoy a few pints of it on the quayside on a summery evening in their native Cornwall.

Verdant say it's the beer that has given them the most trouble. They already had a beer that appealed to the hardcore craft beer bros who love the style: this had to be a crossover beer. It was constantly tweaked with each new brew, until it was just right. As they say on their website, 'When we smash that balance between full-bodied juice and light, zingy cask ale drinkability, it's probably our best beer.' It's sessionable and quenching, and fruity and heavy with hop oils. Unlike many beers in this category, there's also a firm biscuity malt base.

People used to say that modern craft beer styles were incompatible with the traditional British cask ale format. Lightbulb is the beer that proved them wrong.

THE SONG

Delays, 'You Wear the Sun', Rough Trade, 2004, *Alternative Pop*

Delays should have been much bigger than they were. Formed in 2003 by singer and guitarist Greg Gilbert, they had a jangly indie guitar sound. Halfway through recording their debut album, 2004's *Faded Seaside Glamour*, Greg's brother Aaron joined the band on keyboards and backing vocals, and their sound evolved into something lusher.

The centre of Delays though was Greg Gilbert's astonishing voice. His beautiful falsetto drew comparisons with female singers such as Stevie Nicks and the Cocteau Twins' Elizabeth Fraser. He swooped and soared, intimate and next to you one minute, then flying off into the sun the next.

In 2016, Greg Gilbert was diagnosed with inoperable cancer. With the band on hiatus due to family commitments, he launched not one, but two other successful careers, as both a visual artist and a poet. He died in 2021, aged forty-four.

Gilbert described Delay's debut album as 'the sun'. The standout track for me is 'You Wear the Sun'. Lyrically, it's a straightforward ballad about lost love, raised by that lovely, poetic title line. Musically, it begins with a scratchy guitar and builds via a simple four-note keyboard line to a gorgeous sun-drenched symphony that leaves your mind on a quayside, eyes dazzled, sea salt crystalising on your legs.

THE PAIRING

Every time I do my show at Green Man, as the shadows of summer lengthen, I do a pairing of a golden ale or hoppy pale ale with a band playing the festival who have guitar-led songs full of woozy summer charms. I wish I'd been to do this one there, but Delays never played Green Man – at least in the time I've been going there.

Contextually, it's about golden yellow summer sun – fairly easy and obvious, but no less pleasant for that. But there's also a crossmodal element at work here. There's sweetness in a beer like Verdant, and close vocal harmonies and highly melodic tunes evoke sweetness. Gilbert's voice doesn't need anything to harmonise with. It nails the sweetness all on its own.

Kriek Boon with Dizzee Rascal 'Bonkers'

THE BEER

Kriek Boon (4%), *Geuze,* Belgium

Beer is generally bitter or sweet, but there's been a trend for a few years now back to so-called 'sour beers' – styles that were almost forgotten when I started writing about beer.

Yeast is what makes beer, and for the last 140 years, most brewers have used laboratory cultivated yeasts to produce consistent, pleasant beers. But wild yeast is all around us – on our clothes, in our hair, everywhere. Some Belgian beers use wild yeast cultures that give the beer a sharp, complex taste that can never quite be predicted, in a style known as lambic. These beers are then aged in wooden barrels that are home to more bacteria that eat the sugars in the beer and produce dry, funky notes creating an overall character that the word 'sour' does not do justice to. Then, barrel-aged beers are blended with fresh lambic to create geuze. And finally – with this particular beer – 25g of cherries are added to every litre. The sugar ferments out, leaving a tart, fruit taste.

Frank Boon started doing this when the style was almost extinct. Now, his beers win pretty much every category they're entered into in international beer awards. If you don't like the mouth-puckering taste to begin with, it's really worth persevering.

THE SONG

Dizzee Rascal & Armand Van Helden, 'Bonkers', Dirtee Stank Recordings, 2009, *Grime*

Dylan Kwabena Mills was born in 1984 in Bow, East London. While life was tough for young Dylan – his dad died when he was young – he was surrounded by music, with the UK Garage and R&B scene on his doorstep. He was also a big fan of American grunge. Music was his safe space: he was violent and disruptive, and in one school was excluded from every lesson – apart from music. His teachers saw his potential. By the time he was fourteen, he was a drum and bass DJ, and rapping over tracks.

By sixteen, now working under the professional name Dizzee Rascal, he had self-produced his debut single. His first album, 2003's *Boy in da Corner*, won him the Mercury Music Prize, and saw him hailed as one of the pioneers of Grime, a new genre of electronic dance music characterised by rapid breakbeats and jagged electronics, with fast, rapped vocals often talking about the harsh life on London's council estates.

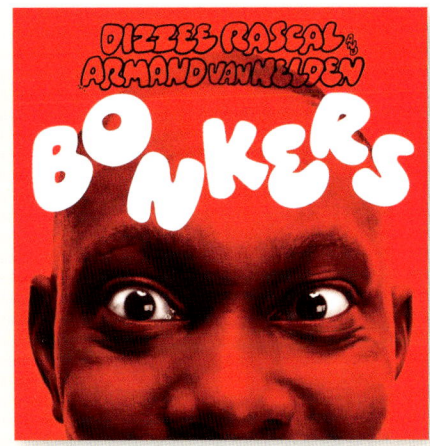

2009's 'Bonkers' was a bit of a departure – an unapologetic festival anthem that blended Grime with influences from House music. It was his second UK number one. The song is a giddy rush, the lyrics a defiant mission statement from Dizzee about living life just as he pleases, once described as 'bubblegum nihilism', over beats that sound like an out-of-control, drunken sugar rush. There's sourness in abundance here – but also a little bit of sweetness.

THE PAIRING

Crossmodal theory states that sour flavours tend to match to music that is high-pitched, dissonant and fast – music that might have the right notes, but possibly in the wrong order. This match is all about dissonance and pitch, unpredictability and strangeness, things not quite being how you expect them. While the personality and tone of the beer and track may seem very different – classic Belgian brewing versus British inner-city Grime – it's about the *feeling* the two have in common. The beer slides around those breakbeats and squelchy synths, like the oil that makes them work.

When I try to pair music with sour beer in live events, it doesn't always work: the degree of sourness and the degree of dissonance have to match. That's why this one works. The beer is quite balanced, and the dissonance isn't too extreme. The music actually mellows out the beer, digging out the hints of sweetness that pair with the music's 'bubblegum rush'.

Athletic Run Wild IPA with The The 'Uncertain Smile'

THE BEER

Athletic, Run Wild (0.5%), *IPA*, USA

It was a challenge by Malcolm Garrett, the designer of the cover for *Tasting Notes*, and also the man who did iconic record sleeves for the likes of Buzzocks, Duran Duran and Simple Minds, to include a non-alcoholic beer.

It was a fair challenge. For a beer to get into this book, it has to have a well-defined character and be pleasant to drink. Thankfully, low- and no-alcohol beers now tick these boxes more often than not.

For breweries like Athletic, they have to. This is not an alcohol-free version of a much-loved full-strength beer. They only make non-alcoholic beers. If these beers weren't good, they wouldn't have a business.

The hop intensity of IPA provides character and flavour that might be lacking in, say, a lager. Here, the hop character is as fulsome as it would be in any IPA, bursting with pine and citrus. In a few short years, the beer has acquired more than enough gold medals in blind tasting competitions to prove its point.

THE SONG

The The, 'Uncertain Smile', Epic/Some Bizarre 1982/1983, *Pop*

This song is so good it single-handedly got Matt Johnson – the man who is The The – a record deal. Johnson is one of the greatest songwriters of his generation. The music blends genres, the lyrics are uncompromising at the same time as being so vivid they could be from musical theatre.

It was released as a single in 1982, then re-recorded for the band's debut album a year later.

I rediscovered the original single recently, and realised this was the version I'd fallen in love with back in the day.

The xylorimba is used more and is higher in the mix, and I love the flute refrain in the chorus and the way the smooth, smoky sax combines with the xylorimba in the solo.

The re-recording for the album is still great. It's much slicker, and the sax is replaced by an extended, jazzy piano solo that may well go down in history as the least irritating thing Jools Holland has ever done. It makes the song. Apparently, Holland was never able to quite repeat it, but he had a stab at it when The The appeared on *Later...* It's definitely worth checking out both versions.

THE PAIRING

In 1997, somewhat foolhardily, I attempted to recreate the journey of India Pale Ale from Burton-on-Trent to Kolkata. That's how I found myself on a container ship for five weeks, with no contact with the rest of the world. There were seventeen Filipino sailors, five German officers, and me. None of them spoke to each other, or to me. I had a breakdown and temporarily lost my sanity.

I did manage to get a bit fitter though. The ship's 'gym' consisted of an unused table tennis table and an exercise bike. I spent about an hour a day on the bike and then did some floor exercises in my cabin. 'Uncertain Smile' had been buried in my music library for years. I dug it out and put it on my gym playlist. Every day, I'd try to work up a sweat by the first time Matt Johnson sang 'if the sweat pours out'. The rhythm was good for pedalling to. The song briefly made me feel as though I was winning, not drowning.

That memory makes 'Uncertain Smile' the perfect match, in my head, for a low-alcohol beer. Athletic beers are a mainstay for me when I take a break from booze, trying yet again to get a bit healthier.

Of course, no one else has this association. Will the pairing work for you without it? I bet it does. How can you not feel virtuous when that piano solo hits?

Stone Smoked Porter with Lift To Experience 'Just as Was Told'

THE BEER

Stone, Smoked Porter (5.9%), *Porter*, USA

Smoked beers pop up a few times in this book because they are so distinctive and work so well in a particular type of pairing. It's also nice to be able to feature a beer from San Diego's Stone Brewing, one of my first introductions to the technicolour world of American craft beer.

Stone were pioneers. The label copy for their flagship IPA, Arrogant Bastard – *'This is an aggressive beer. You probably won't like it. It is quite doubtful that you have the taste or sophistication to be able to appreciate an ale of this quality and depth'* – was plagiarised by British self-styled brewing 'punks' back in the day.

Behind the bluster was great beer. Alaskan Smoked Porter introduced the style to the US in 1988, combining the wood-smoked malt of German rauchbier with the then pretty much extinct English ale style.

Stone's take arrived in 1996, using peat-smoked rather than wood-smoked malt, and it's been around ever since. Peat smoke joins coffee and cocoa on the nose, playing a supporting role rather than dominating. It opens out on the palate with smoke and cocoa, which make it a great breakfast beer. No, seriously – check out the food pairing suggestions on their website.

THE SONG

Lift to Experience, 'Just as Was Told', Bella Union, 2002, *Post-Rock*

All hail post-rock! A genre that emerged, really, in the 1990s, and took flight briefly in the 2000s, it was all about deconstructing the rock song, playing with textures and atmosphere. Sure, there were still guitar solos and stuff. Occasionally there might even be vocals. But it was more sprawling and ambitious than that. This is what prog rock could have been if it hadn't sucked.

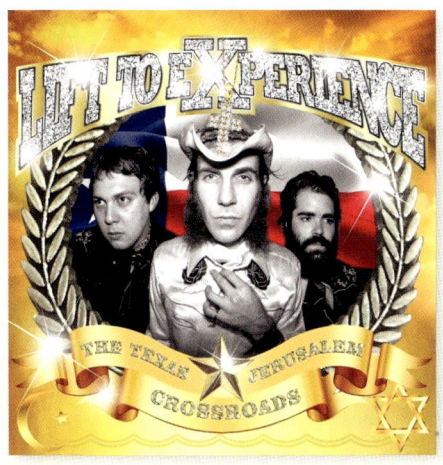

Lift to Experience exploded into the post-rock world in 2002 with a double-concept album about Jesus returning to earth and building a second Jerusalem in Texas. As you do. I think they meant it – I don't think this was ironic – or if it was, it was an example of how the brain can sometimes think two opposing thoughts simultaneously. It's the kind of music God might play if effects pedals and amplifiers had existed in ancient Judea, that's for sure.

The band soon fell apart under the weight of their sheer impossibility. Their founder, guitarist and vocalist Josh T. Pearson later issued an astonishingly honest, savage, intimate and painful solo album which explored mental breakdown and the demons that drove him into it. It's a difficult listen, but worth it. Meanwhile, this opener from what seems likely to be Lift to Experience's only album is like being stuck on a level crossing while a train driven by God and packed with angels wielding rifles screams its way towards you.

THE PAIRING

In 2017, to my absolute astonishment, Lift to Experience not only managed to reform somehow; they played the Green Man Festival. I was so excited, I had to use them in my show. A Welsh smoked porter on the beer list – from a brewery sadly no longer with us – simply seemed like the only possible pairing. It just leapt out. Later, I figured it had something to do with density and weight, but there's more lurking deeper than that.

A tiny, middle-aged Welsh woman came up to me after my event and said, in her lilting valley's accent, 'Well, I don't even drink beer and that made me cry.' The pairing had worked. That made my day. The next day, the same woman spotted me in the crowd gathering for Michael Kiwanuka. Again, she came up to me and tapped me on the arm. This time she said, 'That band, they were doing a signing in the Rough Trade tent. I bought the album and queued to get their autograph, and I told them all about you. And they had no idea who you were!'

Utopian × Gadds' Green-Hopped Pilsner with Ralph Vaughan Williams 'The Lark Ascending'

THE BEER

Utopian × **Gadds'**, Green-Hopped Pilsner (5%), *Lager*, UK

Hops are fragile things. As soon as they're picked from the bine, they start to fade. This is why they're kilned and dried within minutes of harvesting: the process means we lose some of their rich bouquet, but preserves what's left for much longer.

The only alternative to this tough compromise is to brew with the hops as soon as they're picked, when they're still 'green' or 'wet'. Kentish brewers are well-positioned to do this, being closer to the hop gardens than most, and Ramsgate's Eddie Gadd is a master of the green hop. Utopian combine lager brewing excellence with a commitment to only brewing with British ingredients. So this collaboration between the two, featuring fresh East Kent Goldings hops, is a compelling proposition.

I'd always thought green hops would give more intense spikes of hop character, but that's not the case. Instead, they give a broader, more colourful palate of flavour, richer and sappier. And, yes, I suppose, *greener*.

THE SONG

Ralph Vaughan Williams, 'The Lark Ascending', (This recording by Nicola Benedetti and the London Philharmonic Orchestra, Decca, 2012), *Classical*

Classical music can sometimes feel elitist because it requires a level of knowledge of the form to fully appreciate what's happening. In *This is Your Brain on Music*, Daneil Levitin describes two parallel experiences of listening to a Gustav Mahler symphony. The knowledgeable person goes 'Ah, I see what he did there, that's really clever, subverting the conventions of the form in that way,' while the philistine just hears a noise.

The Spotify liner notes for Ralph Vaughan Williams talk about how in an earlier work, 'A Theme of Thomas Tallis', the composer 'introduced antiphonal effects within the context of modal tonality, juxtaposing consonant, but unrelated triads.' Mm-hmm. Whereas the reason 'The Lark Ascending' is one of the most widely popular pieces of classical music in the world is that, whatever he might be doing here with

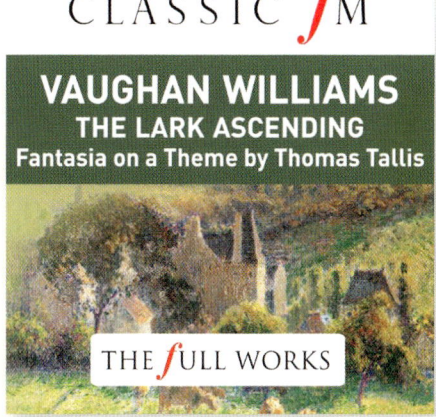

triads and modal tonality, the music does something any listener can understand. The orchestra represents a rural British landscape. The violin soars above it, describing both the song and the flight of the titular lark. It's as simple to understand as it is incredibly beautiful. Here, music creates shape and movement in the minds of anyone who hears it. There's more that can be discussed without losing the layperson. Does it reflect Vaughan Williams's interest in English folk music – can you hear any of that in there or not? He composed it in 1914, before going off to be a medic in the trenches of the First World War, but didn't record it until after the war had ended. Is it a straightforward evocation of the English landscape? Or does it contain hints of nostalgia for a lost pre-war innocence? Is the singing lark calling out to a lost mate, or simply exulting in the joy of a spring morning? Dunno. Might have to listen to it again. And again.

THE PAIRING

As contextual pairings go, you can't get much better than this one: the English landscape evoked as sound, and as flavour. Vaughan Williams's lark could easily be flying above the bines and oast houses of Kent. The fragility and delicacy of the lark is echoed by the fleeting glory of the hops themselves. The verdant green of the can runs like paint into the music. Green hop beers, by their nature, tend to appear in the autumn, harvest time giving up the fruits of the summer's warmth. Utopian and Gadd's have done this collaboration a couple of times now and it's likely they will again. I'll be chilling a few cans and waiting for spring, when 'The Lark Ascending' soundtracks the return of pleasantly warm days, and you find one such day that taps into the song's nostalgia for a time before everything seemed so scary.

Anspach & Hobday The IPA with The Blue Nile 'Tinseltown in the Rain'

THE BEER

Anspach & Hobday, The IPA (6%), *IPA*, UK

Anspach & Hobday sounds like a pastiche of a Victorian brewing company. While there are elements of such pastiche in their branding, these are the real surnames of the men who founded the brewery in 2013. Their motto is 'Traditional beers brewed in a modern way', and this approach serves them well. Each beer feels considered, pondered over, refined until it's just right.

A&H are now best known for London Black, their cheeky challenge to the dominance of Guinness in porter and stout. But for me, the IPA is their triumph. This is a beer style that evolves constantly. What is now considered the norm for an IPA is in many ways the opposite of what the style was just twenty years ago. Here, A&H defy convention and have a go at how we think IPA was in the mid-nineteenth century – but with that all-important modern twist in the guise of new hop varieties.

The result tastes like a cross between a traditional British IPA and something from the first wave of North American craft brewing at the end of the twentieth century. The hops are rich and aromatic, fresh and urgent. There's a lot of hop bitterness at the end to counter the heady aromas up front. But they're backed by a deep layer of sweet caramel and digestive biscuit malt character. We often say of drinkable beers that they belie their strength. Here, you know you're drinking a strong beer from the first sip. But you'll still go back for a second pint.

The Blue Nile, 'Tinseltown in the Rain',
Linn/A&M, 1984, *Pop*

In the first half of the 1980s, something wonderful happened in the Glasgow music scene. A city that was constantly being depicted on TV as a semi-derelict, drug-and-crime-addled wasteland simply decided that it was the coolest, most easterly city of the United States, and began to spit out band after band who purveyed a slick combination of soul, pop, R&B and indie. It came with smart clothes and sharp attitudes. It was glossy and refined. One journalist labelled it 'sophistipop', a label which was, mercifully, quickly forgotten.

The Blue Nile were the best of the lot. Absolute perfectionists, they managed to record a grand total of four albums in their twenty-three-year career. At least three of these albums are masterpieces. 'Tinseltown' was the second single from their debut, *A Walk Across the Rooftops*. It peaked at number 87 in the UK charts, but continues to influence artists down the decades. Even Taylor Swift is a big fan. It's an ode to Glasgow, reimagining the city as New York. It's a stirring anthem to the redemptive power of love. It's the standout song from a standout album that really should have been turned into a West End musical by now. Just magnificent.

THE PAIRING

This pairing is about bitterness. As a beer fan, I was always surprised in crossmodal experiments that bitterness was perceived as unpleasant. Bitterness pairs with music that is low in pitch and legato – the notes are smooth and connected, with no gaps or silences in between. That's what we get here with The Blue Nile – a smooth, funky bassline under deep, swooping synths and dark chocolate-like strings. Paul Buchanan's voice over the top provides a balancing sweetness, but the song as a whole has an assertive bitterness that pairs with the uncompromising finish of this IPA. It all sounds quite challenging and aggressive on paper. But taste the beer, listen to the track, and you'll struggle to understand why people think bitterness in beer is a bad thing. And why the song only got to No. 87 in the charts.

Viven Smoked Porter with The Unthanks 'Magpie'

THE BEER

Viven, Smoked Porter (7%), *Porter*, Belgium

'Smoke, chocolate, and British bravado in a single glass. Dark as night, smoky as an old pub, this is a beer that melts on your tongue like chocolate.'

I have to admit, I'm quite envious of this copy from Viven's website, and thought it worth quoting here. It reveals that, while British beer aficionados undoubtedly adore Belgian beers, the relationship works both ways.

When you first encounter the wide, wild world of Belgian beer styles, it feels completely different from anything you've seen in beer before. But look at it more closely and relationships emerge. Like the UK, most beers drunk in Belgium now are brewed by multinational corporations. (At least Belgium can claim Anheuser-Busch InBev as a Belgian rather than foreign company.) But also like Britain, there's a strong ale-brewing tradition that never died out under the lager onslaught. Sometimes the similarities are buried under layers of innovation and experimentation. Here, it's a simple Belgian tribute to a classic British beer – with just a slight twist.

Viven started brewing in the tiny Belgian village of Vivenkapelle, just outside Bruges, in 1999. In 2005, new owner Tony Traen expanded from the traditional Belgian blonde and brune beers that formed the core range, first brewing a traditional porter and IPA, and then a smoked porter.

The smoked version is quite bitter, with roasted mocha and chocolate elements tempering the smokiness, and a lingering finish. It's one for contemplation.

THE SONG

The Unthanks, 'Magpie', RabbleRouser Music, 2015, *Folk*

If I'm well-prepared, I can listen to the Unthanks just fine. I'm OK. But if they creep up on me unawares, on the soundtrack to a TV programme or on the radio, I burst into tears. I can't help it. Rachel and Becky Unthank are sisters, born seven years apart. Rachel's vocals are crisp and clear, like a ringing bell made from crystal glass. Becky's, by contrast, are smoky and soft, sometimes more sighing breath than singing. In harmony, these two voices create something hauntingly

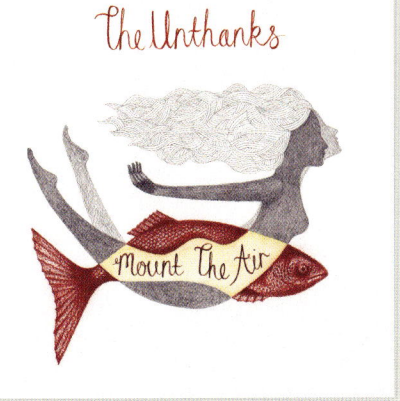

beautiful. They're very much in the folk tradition. But they transcend the genre.

The song 'Magpie' is a version of an ancient nursery rhyme that honours a bird steeped in superstition and folklore. It was used at the end of the first episode of the third series of the BBC show *Detectorists*. Bastards caught me off-guard.

THE PAIRING

I first paired this song with a porter at the Green Man Festival, when The Unthanks were playing. I shared the pairing on social media over the festival weekend, and the band actually replied, saying they love a good porter. So I guess I got this one right.

Primarily it's about pitch: low, husky-pitched vocal and backing paired with smoke, deep chocolate and coffee, which have all been proven to correlate with low musical pitch. It's also, in particular, about pairing the smoky character of the beer with Becky Unthank's voice. Smoke itself is completely silent, so I'm not quite sure what we mean when we describe a voice like that as smoky. But some kind of crossmodal relationship means it definitely is. It's about tone and the subject matter too. There's something about the mystery and mood of the song that begs to be paired with a dark, contemplative beer.

Finally, there's also a nice contrast in there between the city-based origins of porter – the beer of the industrial revolution, with its grease and machinery and railways – and the bucolic fields over which the magpie keeps watch. Ancient and modern, mingling in the darkness.

Stone & Wood Pacific Ale with Tim Buckley 'Buzzin' Fly'

THE BEER

Stone & Wood, Pacific Ale (4.4%), *Pale Ale*, Australia

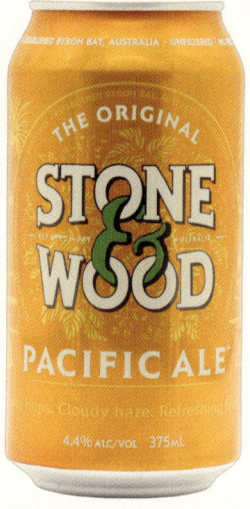

The three founders of Stone & Wood were veterans of big brewing who wanted to do something different. They sat in the sun outside the Beach Hotel in Byron Bay, Australia, watching the surfers come in. Imagine you were walking up the beach to this palm-covered bar, board under your arm, the sea salt drying on your legs, they challenged each other. What beer would you want to drink right at that moment? They agreed that the perfect beer would be a slightly hazy pale ale, a showcase for Tasmania's much-coveted Galaxy hop, with its intense peach and passion fruit aromas. They brewed this beer, called it Pacific Ale, and launched one of Australia's most successful craft breweries off the back of it.

While there are some stylistic similarities to American hazy pales, they were right to declare that Pacific Ale wasn't just a name; it was a new beer style. The flavours are not New England flavours; they're Australian flavours, quite specifically.

In 2016 I visited the hop harvest in Tasmania. We stood on a hill overlooking the farm, and waves of peach aromas drifted towards us on the breeze. I then took a box of green, wet hops on an interesting flight from Tasmania to Byron to get them into muslin bags inside the fermenters within 24 hours of being picked. I didn't get to taste that batch until later. But while the hops were steeping, I did get to sit in the sun at the Beachie, drinking Pacific Ale without having to go to the trouble of surfing beforehand. The Beachie sits right on the sand, essentially a huge, semi-covered courtyard. Many of the beer fonts are encased in ice. Pacific Ale is, emphatically, the flavour of Byron Bay, which you can now enjoy in grey Britain too.

THE SONG

Tim Buckley, 'Buzzin' Fly', Elektra, *Jazz-Folk*, 1969

Having died in 1975 when he was just twenty-eight, Tim Buckley was arguably more famous in the late eighties and early nineties than when he was alive. He wrote 'Song to the Siren', which has been covered countless times and appears elsewhere in this book in its best version. From that song, I went back to explore his albums when they were reissued on CD in the 1990s. Some of it is too folky for me. The experimental stuff is a bit too rich for my tastes. But his extraordinary five-octave voice soars constantly through his work.

'Buzzin' Fly' is the song that made the less agreeable stuff worth persevering with. It's one of the happiest, most joyful songs I've ever heard. It's about sun-drenched new love, and even though some of the lyrics suggest that love is now over, this hasn't dimmed the delight of meeting someone and declaring that you want to know everything about them, as seabirds call their name.

Buckley is the buzzin' fly, his voice and guitar completely wrapped around each other, like two sides of the same instrument, or like he's singing with both.

THE PAIRING

This is a pairing about summer, or more specifically, about the sun itself. The guitar sounds like sunlight. The beer looks and tastes like liquid sunlight. Given that the beer was conceived so specifically for a perfect sunny moment, it was always going to be open for a contextual pairing in this vein. 'Buzzin' Fly' should be on constant rotation at the Beach Hotel in Byron Bay.

The crossmodal part is a correspondence based on sweetness. Here, both Buckley's soaring voice and joyful guitar playing are high-pitched and melodic, which means they pair perfectly with the sweet peach and passion fruit of the galaxy hops.

Signature Brew Black Vinyl with Etta James 'At Last!'

THE BEER

Signature Brew, Black Vinyl (4.5%), *Nitro Stout*, UK

Signature Brew are soul brothers in my quest to pair beer and music. They're so into the idea, they started a brewery in 2011, with the aim of providing good beer in music venues. Core beers now have names like 'Roadie' and 'Backstage IPA'. But this is far more than just a gimmick. Their very first beer was a collaboration with London indie-rock band The Rifles. They now have two sites in London, both providing regular gigs in a city that is losing many of its live music venues. In lockdown, Signature Brew created a 'pub in a box' for home delivery, and hired out-of-work musicians to do the deliveries.

Black Vinyl, then, is the perfect name for a smooth, slick, nitro stout. Brewed as a one-off, it was so popular it became part of the permanent core range. It's one of those stouts that's like a grown-up Guinness. Everything you want from dark, roasty malt is in there: chocolate, a hint of coffee, an even fainter, just detectable hint of tobacco, all given a little bit of lightness by a few more fruity American hops than you might expect. But this is all wrapped up in a lush, silky mouthfeel that makes it all very mellow and drinkable, allowing it to finish dry, but not at all astringent.

THE SONG

Etta James, 'At Last!', MCA/Chess, 1960, *R&B*

Etta James's remarkably powerful voice started showing when she was just five years old. The adults around her loved that voice, and inflicted awful violence on her because of it. Born in 1938, she never knew her father, and as a young girl hardly saw her mother either. She was passed around between friends and relatives who did their best to bring her up. Soon, she was a soloist in the local church choir. The choir's musical director punched her in the chest to force her voice to come from her gut. Her main foster father would wake her up at night and beat her to scare her into singing for his friends.

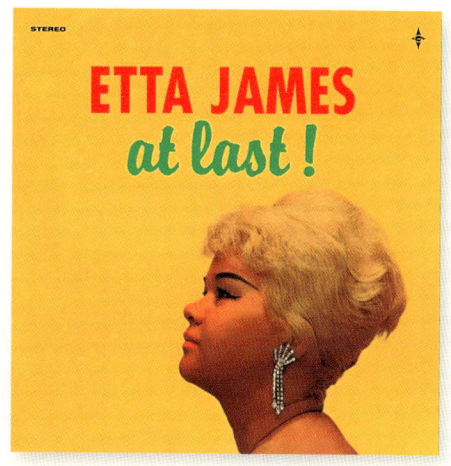

By fourteen she was singing doo-wop and blues in nightclubs. By sixteen she was in a relationship with B. B. King. She's still regarded as one of the greatest blues singers of all time. But in 1960, the Chess brothers, founders of the legendary Chess records, spotted a crossover appeal in her. For her debut album, they surrounded her with lush strings, and chose a collection of genre-spanning songs that showed off not just the power, but also the richness of her voice.

'At Last' would become her signature tune. There's something quite wonderful about taking the rugged, ragged emotion of the blues, and applying it to music that is *happy*. The album has an opulent, rich-as-in-wealthy sound, and James sings like a queen saying 'fuck you' to everyone who abused her on the way up.

It's a sound that's become a mini-genre in its own right: the sound of TV ads for aspirational lifestyles. 'At Last' has long been a standard at weddings and receptions, and has been used on ads for Cadillac, Jaguar, Porsche, and more recently, Sainsbury's and, er, Heinz ketchup.

THE PAIRING

As I established earlier, I'm not above stealing, and I stole this pairing very happily from Signature Brew themselves. On their website, they have a short playlist of about ten songs for each beer in their core range. Most of these songs are silky and smooth, many specifically mentioning chocolate or coffee in the lyrics. This one doesn't. But both the beer and the song are smooth and luxurious. This pairing is like that moment when you stretch out on a sofa with a mug of coffee, a bar of smooth chocolate, or both. Maybe because I've been conditioned by so many of those TV ads.

Duvel with Pixies 'Debaser'

THE BEER

Duvel (8.5%), *Golden Ale*, Belgium

Never let anyone tell you Belgium is boring. Only someone who doesn't understand Belgium and has never been there would recite the tired old challenge to name ten famous Belgians. It's easy: Westmalle, Chimay, Orval, Saison Dupont... There are many more. And firmly near the top of the list is Duvel.

After the end of the First World War, the Moortgat brewery wanted to create a beer to celebrate the contribution made by British troops. Albert Moortgat travelled to Britain in search of a yeast to recreate the character of Scotch ales, and found what he was looking for in a bottle of McEwans. He cultured the yeast from the sediment in the bottle, and created a strong, dark beer which he named Victory Ale.

In 1970, the ever-increasing popularity of Pilsner prompted Moortgat to re-engineer their flagship beer. They wanted to keep the oomph, but change the appearance so it looked like a lager. They managed to make Pilsner malt and that old Scottish yeast work hard enough to yield an 8.5% beer that prompted one brewer to exclaim, 'This is a devil of a beer,' and thus, Victory Ale became Duvel.

Duvel certainly packs a punch, but it tastes smooth and light, although a sixth sense tells you there's some power in there somewhere. Hints of lemon, fresh bread and a touch of spice mean this beer does a very good cover version of what white wine normally does.

THE SONG

Pixies, 'Debaser', 4AD, 1989, *Alternative Rock*

It's not often that you hear something that truly – honestly – blows your mind. For it to have an effect so dramatic that you remember it for years to come, it has to sound utterly unlike anything you've heard before. You love it, but you're also a little frightened by it.

This is what Nirvana were listening to when they were developing their sound, and you can hear all the elements that later became defined as 'grunge'. But then, over the top of the rangy bass and scratchy guitars, comes the voice of Black Francis, a man possessed, screaming about slicing up eyeballs and Andalusian dogs. I once found a recording of that incredible vocal, isolated, frenzied and manic. It sounds

like a lunatic preacher ranting in the street, completely losing it, shredding his vocal chords. This is a song of extraordinary power that swaggers into the room and slaps you in the face to get your full attention.

THE PAIRING

This is one of the very first pairings I did, and is still among the most successful. It just works on so many different levels.

On the first approach, they're both quite visceral: the intro to the song – the first track on 1989's *Doolittle* – feels like a physical assault. Meanwhile, pour Duvel into its trademark tulip glass and even if you're a skilled pourer, the foam surges aggressively, as if it's trying to get out of the glass and punch you in the face. Both the beer and the track come at you in an intoxicating, thrilling attack.

On a deeper level, Belgium was one of the most important countries in the artistic movement of surrealism. If you've been to Belgium, this makes a lot of sense: either surrealism directly influenced the evolution of Belgian beer, or vice versa, or both.

What does this have to do with 'Debaser?' Well, it's lyrics make explicit reference to the 1929 surrealist film *Un Chien Andalou*, by Luis Buñuel and Salvador Dalí. Famously, the film contains a scene in which a man seemingly slices a woman's eyeball with a scalpel (it was actually done with a clever cut to a cow's eyeball) and Francis's extraordinary vocal performance here can be read as a reaction of thrilled disgust at having just seen it. You'd need a beer like Duvel after that.

Among my live event audience, this pairing often creates a Pavlovian response: later, when people hear the song, they crave the beer. Or when drinking the beer, that menacing bassline appears in their heads. Whoa-ha-ha-hao!

Ampersand Experiments in Evil with Neutrinos 'Heaven'

THE BEER

Ampersand, Experiments in Evil
(10–11%), *Imperial Stout*, UK

We beer fans can be very predictable. Whisper a rumour of a rare, barrel-aged imperial stout, and we salivate.

Ampersand was founded in 2017 by three siblings on their Norfolk family farm. They expanded to fill a new brewery in the town of Diss, near the border with Suffolk, in 2021. Every year, they brew Experiments in Evil, ready to launch at the beginning of December. The experimental part is that each brew matures in a barrel with a different history. In 2024 there were two: the 8th edition was aged in Islay whisky barrels for eighteen months, while the second was aged in rum barrels with Brettanomyces yeast for two years.

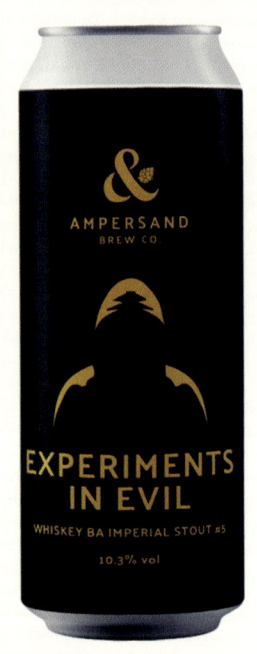

I moved to Norwich in December 2023, and Experiments in Evil #7 – aged in brandy barrels for five months – had just hit the shelves of Sir Toby's, a bottle shop and bar perched on the edge of Norwich market. It was the first beer I bought in my new home town. I sat watching people do their Christmas shopping while I savoured the warming smoothness, vanilla notes, roasted coffee and dark chocolate, amazed at its drinkability, and thought, 'Welcome home'.

THE SONG

Neutrinos, 'Heaven', Wetnurse Records, 2022, *Alternative Rock*

The Neutrinos are a band that formed in Norwich in the late 1990s. A year after moving to the city, Liz told me we were going to see them. We had to. OK, what venue were they playing? Their house.

Many bands gather a loyal local following and stacks of critical acclaim without ever breaking through on a national scale. When The Neutrinos didn't make it in the conventional sense, they decided to abandon convention, and reinvent what a live gig could be.

Over the last few years, this has centred on KlangHaus, a music/art installation that takes place in the old city centre house that several band members call home. It's an intimate, multi-media affair that breaks down

the barriers between band and audience. Performers move around, and encourage the audience to do so too. Which is how I found myself standing in an old fireplace, a grand piano to my right, a drumkit to my left, as they played their final song. I was so close, I could feel the music physically. I felt naked and exposed as it built and built to a squall of lovely noise, and I simply couldn't cope. Soon we were filing out, past merch and comment cards and smiling band members, and I had to get out into the street without making eye contact so I could collapse and sob. It filled me with emotions, then emptied me out, like one of the intense rainstorms when there's hot sun and blue skies ten minutes later.

The Neutrinos have recorded 'Heaven' several times. There are two versions. They're both wonderful, but they're nowhere close to the music they played in the house, while I was standing amid the instruments. Nothing else can be.

THE PAIRING

When the Neutrinos finally released 'Heaven' as a single in 2022, they said, 'This song has lived with us for decades and has been so challenging to record, too sweet, too bitter, too fast, too slow and finally we have learned how to just let it go.'

You'll have spotted the crossmodal association they slipped in there. There is bitterness and sweetness in the song, as well as sourness and maybe even a hint of saltiness. But there's more to crossmodal pairing than pitch and basic tastes. For me, this pairing is about the idea of *weight*. The song and the beer are both heavy. They're weighted down by layers, but not in a bad way – more in the way that a heavy duvet feels comforting in the heart of winter.

For me, they go together as my welcome to the beer and music of Norwich. It's also notable that both the song and the beer are not fixed, but constantly evolving. But it's the feeling of being weighed down by comforting layers that really wraps them together.

Allsopp's Arctic Ale with Fred Again.. and Brian Eno 'Enough'

THE BEER

Allsopp's, Arctic Ale (11%), *Barley Wine*, UK

Samuel Allsopp was the first brewer in Burton-on-Trent to brew India Pale Ale (IPA). His brewery had a reputation for brewing beers that could survive long voyages, first established by brewing Burton Ale for the Russian court in St Petersburg. So in 1851, when the British Admiralty asked for a beer to supply ships sailing to the Arctic, it was of little surprise that Allsopp's won the contract.

The Northwest Passage was a shipping route across the north of Canada that existed only in theory. It was an alluring concept though – potentially a shorter route to the Pacific than the long, perilous trip around Cape Horn, long before the foundation of the Panama Canal. The first ships sent to explore the route, in 1845, disappeared. In 1852, a new expedition set off to see what had happened to them. This trip was equipped with a strong ale that wouldn't freeze anywhere above 40 degrees Celsius. It was described as rich, vinous and nutty, and aged well, resembling old Madeira.

Samuel Allsopp's seven-times-great-grandson, Jamie, revived Samuel Allsopp & Sons as a brewing concern in 2021, and recreated Arctic Ale from an 1875 recipe. I guess you could say it's stunning in more ways than one. I judged it blind in a beer festival of heritage beer styles, and it won its category. I was delighted to find out it was this beer, and that its current incarnation is well worthy of one of beer's greatest legends – potent and deep, warming and brooding. The last brew was a one-off. As I write, there's one cask of it left in the cellar of Jamie's pub, the Blue Stoops in West London. But he promises me there will be a new brew fermenting by the time you read this.

Fred Again.. and Brian Eno, 'Enough',
Opal, 2023, *Ambient*

Brian Eno has a choir that convenes in his north-west London house. Frederick John Philip Gibson was the sixteen-year-old son of Eno's neighbours when he joined the group. Seven years later, Eno, arguably the most influential musician and producer of the last fifty years, was crediting Gibson, now recording and producing as Fred Again.., with teaching him about contemporary music. 'I didn't really understand a lot of what he was doing,' said Eno, the pioneering musical genius.

Their collaborative album, *Secret Life*, confused many of Fred Again..'s clubbier fans. But it still carries his trademark dense sonic collages. These squish together our ideas of what sampling and cover versions are, with an additional countless number of layers. This is what Eno found so original.

'Enough' is the standout song on an exceptional album. Somewhere in there, like vegetation crushed to coal under the weight of what's on top of it, are the remains of a song called 'Don't You Dare', by Danish singer-songwriter Winnie Raeder. That forms the foundation for a delicate web of crystalline beauty that takes the breath away.

THE PAIRING

This is another one that I forced a little, because I was desperate to have the song and the beer in this book. But when these two were left staring at each other awkwardly, because they're both different from everyone else and not as easy to get to know properly, I realised they belonged together, their connection as powerful as any other.

We could talk about the sweetness and bitterness coiling around each other. We could repeat the idea that unites many of the strong beers in this book, about corresponding layers and density and depth. These ideas run through different pairings because they are real and true. But what sets this one apart is that evocation of cold and ice. The song sounds frozen. But like the beer, it's still liquid, still moves. I once took Liz for a holiday at the Ice Hotel in northern Sweden. Neither the beer nor the song existed then. I'd love to take them both back there.

Verzet Oud Bruin with
Public Enemy 'Rebel Without a Pause'

THE BEER

Verzet, Oud Bruin (6%),
Flanders Red/Brown Ale, Belgium

*'If we were a band, we wouldn't want to play covers.
We'd want to perform our own stuff.'*

When the founders of Verzet told me this, I was hooked. I'd just
asked them why they'd brewed a beer style that was traditional
and specific to their native Flanders, while every other young
start-up brewery in Belgium seemed to be doing that thing
that every generation does: rejecting what their dads drank,
no matter how good or bad that was, and embracing something
contemporary and usually imported – in this case, hazy
American-style IPAs.

Hazies were what these lads enjoyed drinking themselves.
But brewing that style would have cast them, in their minds,
as nothing more than a tribute act. Here were like-minded
people who saw deep connections between beer and music.
The oak barrels that age beers like Oud Bruin from anywhere
between eight and 24 months each have names: Keith Richards
should perhaps be nervous – he's at the end of a row containing
deceased icons Chuck Berry, Freddy Mercury and Tupac. These
barrels are blended to a pleasing whole – imagine if you could do
that with the musicians! – which is then aged in bottle for another six months.

Verzet collected and cultivated their yeast strain for this beer from outside their
front doors, making it, literally, the taste of Flanders. True to style, it announces
itself with a screech of sourness on the palate. Once that clears, you get red fruit,
red wine, green apple, bread, wood character – oh, and more sourness. But it's
never monotone – there's so much going on.

Public Enemy, 'Rebel Without a Pause', Def Jam, 1987, *Hip Hop*

I don't listen to a lot of hip-hop. But I remember this song, in 1988, barging into the room and dragging me out of my twee indie complacency like a raid.

The sax that screams like a stovetop kettle about to explode was sampled from the 'The Grunt', a little-known track by the J.B.s, James Brown's backing band. Funky though that track is, Public Enemy chose to sample the beat from James Brown's 'Funky Drummer' instead (not for the first time). Sampling was in its infancy, and that improvised drum break by Clyde Stubblefield would go onto be one of the most sampled pieces of music in history.

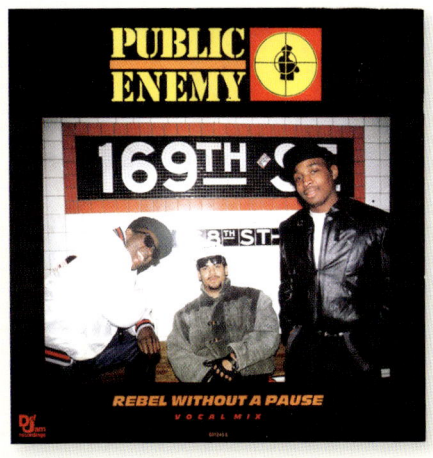

Over a bed of groovy, funky samples, sometimes sparse but always heavy, the group's leader, Chuck D, declares war, pausing every now and again to announce Terminator X, whose scratching adds another rich layer to the assault.

In 1986, Public Enemy's first album, *Yo! Bum Rush the Show!* had placed the group firmly at the centre of political hip hop, but was ignored by white rock critics and radio stations. Just a year later, Public Enemy felt that album was dated in terms of sampling and production technology. There was to be no ignoring this follow-up. It's faster, heavier, denser and thicker. It still feels as vital almost forty years later: something is happening. You need to pay attention.

THE PAIRING

The brewery's own tasting notes for the beer say that it 'awakens your taste buds with a massive wallop'. Just as with Duvel and Debaser, that's something the song and the beer have in common here. The sampled screech pairs perfectly with the sour hit in the beer. They accentuate each other. But while this is a dominant element, there are layers or richness surrounding both. The sourness and the screech would be too one-dimensional on their own. Each works perfectly, standing out from a lot of other detail that's going on around them.

Burning Sky Choose Any Memory with Beach Boys 'God Only Knows'

THE BEER

Burning Sky, Choose Any Memory (5.6%), *Mixed Fermentation Fruited Beer,* UK

The beer industry and its acolytes seem to have moved on now from discussions on the nature and definition of craft beer. In some ways this is a shame. Craft beer ended up being a jaded, exploited term, and that's probably how it will be remembered.

This does a great disservice to brewers like Mark Tranter, founder of Burning Sky. Craft beer was revolutionary because it was a romantic idea that appealed to people who were bored or angry with a monotone, corporate world. Burning Sky is the kind of brewer they probably imagined. Small and farm-based, deep in the countryside, Tranter is here to brew the kinds of beers he wants to brew, not to grow the brewery and sell it.

There are the beers he brews because he knows what will sell, and they keep the lights on. Then there are the beers he tinkers with and obsesses over, taking Belgian traditions and shaping them for the Sussex Downs.

Choose Any Memory is a mixed fermentation beer that has been aged for seven months in both oak barriques and stainless-steel tanks, on fresh raspberries. It sports a crisp acidity and delicate funk from the yeast, which complements and rounds out the fruit character. At the time of writing, it costs over fifteen quid for a 750ml bottle. Not because it's over-priced and over-hyped, but because that's what it's worth – when you can actually find it.

The Beach Boys, 'God Only Knows', Capitol, 1966,
Baroque Pop

Cast your mind back to an age before streaming. Music, when you buy it, is a physical artefact. As a fan, your entire life has been defined by scarcity. You read about classic albums from the 1960s and 1970s, and one day you rifle through your parents' vinyl collection hoping they will have classics that you've heard about so many times, but never actually *heard*.

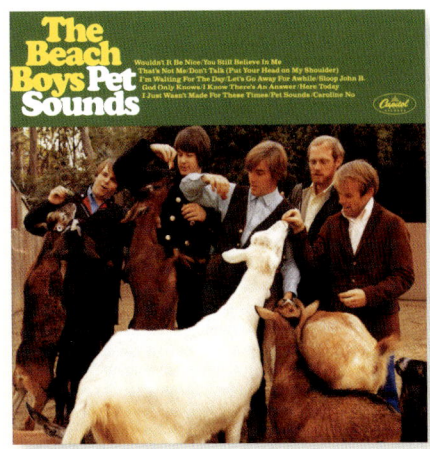

Then, as CDs become commoditised, record companies begin re-releasing entire back catalogues on this new format. You're a few years into your first proper job and you have money to spend for the first time, and they're selling any three new/old CDs for £21. Almost shaking, you buy The Velvet Underground & Nico. You buy artists that were big influences on the bands you love, like Nick Drake, Tim Buckley, Kraftwerk, Bowie's late seventies stuff, Eno, Magazine, Billie Holiday... and *this*.

You always thought the Beach Boys were the poppiest of pop, and never understood why so many people insisted *Pet Sounds* was essential. And then you hear 'God Only Knows' for the first time in your life, when it's thirty years old and you aren't, yet, and your comprehension of popular music spins on its axis. After three plays, you start to dream it, waking up to it playing in your head. Your understanding of both love and genius just got a bit deeper. It's OK: the same thing happened to Paul McCartney when he first heard it too.

THE PAIRING

The Beach Boys' vocal harmonies are my first go-to when I'm looking to pair something sweet, simple and straightforward. But this beer isn't just sweet. There's acidity in there too. The song takes sounds and stylings that we associate with happiness – summer sun and surfing – and uses them to create something sad. There's sweetness and acidity wrapped around each other in the music, and there is in the beer. Each has simple roots, but turns into something more complex and thoughtful.

Heineken with U2 'Running to Stand Still'

THE BEER

Heineken (5%), *Lager*, Netherlands

Heineken is a big corporation, and behaves like one.
But those who think big global brewers are 'all the same'
haven't looked closely enough.

Heineken is still family-owned. As the global head
brewer once told me, 'It's their name on the bottle. If you're
doing something to their beer, they want to know about it.'
Heineken is only brewed with malt, never cheaper adjuncts.
It's lagered for 21 days. I once tasted Heineken fresh from
the fermenters, before filtering and pasteurisation.
It really was a very good beer indeed.

For many palates, including my own, Heineken is a
little sweet, and that makes it easy to dismiss as 'crap'.
But that flavour profile is intentional — and possibly unique.
In most beers, the aroma comes from a late addition of hops.
Heineken has no late hopping. All its aroma comes from
compounds known as esters, which are created by yeast
during fermentation. Estery notes are common in many beers,
especially ales. But I know of no other beer that relies solely
on the yeast for its aroma. That at least makes it interesting.

THE SONG

U2, 'Running to Stand Still', Island Records, 1987, *Rock*

You don't need me to tell you anything about *The Joshua Tree*. It's such a behemoth,
it's hard to remember what life was like before it came out. For millions of music
lovers now, it's just always been there, a standing stone in the cultural landscape
since before they were born.

In 1987, I was in my first year at university. U2 were commercially successful, but still very much an 'alternative' band, added by people like me to compilation cassettes alongside The Smiths, New Order and Echo and the Bunnymen. I bought *The Joshua Tree* when it came out in March.

St Andrews didn't have many alternative indie kids, and new music took a while to get there. When we got back for my second year in October, the sound of this album came out of every single room in my hall of residence. Every pub busker – and there were lots – alternated between songs from this and The Proclaimers' first album. It was everywhere. Inescapable. And it has been ever since. I'm sick of the sound of *The Joshua Tree* now. But 'Running to Stand Still' is by far the best song on it. Still arresting, still moving.

THE PAIRING

I'm not a natural comedian. If I see a cheap joke that's going to get laughs, I have no pride. I just go for it, the goal-hanger who waits for the rare occasions when they can just blast the ball into an empty net.

When I started pairing songs and beers, there were obvious gags that were too juicy to ignore. Ed Sheeran pairs with Corona – music for people who don't like music, with beer for people who don't like beer. Coldplay and a can of Foster's, but not any can of Fosters – the can you wake up next to after an all-night house party, take a swig from, and realise people have been using it as an ashtray while you were out.

At my gigs, I say U2 and The Joshua Tree is a great pairing because both once mattered and stood for something, but long ago sold out on any principles in pursuit of pure capitalist greed. The fact that this is not accurate, and doesn't stand up under scrutiny, has to be weighed against the fact that it's one of the biggest laughs I get, sometimes even with a round of applause.

Both have their place, and for me that place is airside at Abu Dhabi airport at 3.30am local time. After a thirteen-hour flight from Melbourne, during a three-hour layover before another eleven-hour flight to London, in a bar lined with ice-encased Heineken fonts and nothing else, this is the best beer and the best song in the world. But only there, only then.

Westmalle Tripel with Lady Gaga 'Bad Romance'

THE BEER

Westmalle, Tripel (9.5%), *Tripel*, Belgium

Westmalle was the first Belgian Trappist abbey to revive – or rather reinvent – the monastic brewing tradition in Belgium, in 1836. In 1934 they launched Westmalle Tripel, inventing the style.

Tripels are over 8% ABV. This one is 9.5%, making it one of those beers that friends and family sometimes express concern over when I'm drinking one, as they order their second glass of 14% ABV Shiraz. On the surface it's a simple beer – pale gold and clear, moderately bitter and quite fruity. It doesn't drink like a 9.5% beer. It's brewed with Candi sugar, which gives a leg-up to the ABV and also adds another layer of flavour. The yeast esters add a distinct banana aroma, which I would say dominates, except there's also green apple and freshly baked bread. Some of the world's best hop varieties contribute a balancing bitterness, and there's spice and biscuit and pepper, and something new with each sip, which is why you don't drink it like you would a pint, and why there's nothing wrong with a 9.5% beer if you're sipping 330ml of it over the course of an hour or so. It's also bottle-conditioned, which means you'll often get some slight variation in the character as it ages.

Westmalle don't enter international brewing competitions. They don't need to. And anyway, it doesn't feel like a very Trappist thing to do. This means that more than once, I've been sitting at a judging table, with ten or twenty tripels in front of me thinking, 'None of them are Westmalle. So what's the point? If we crown one of these as the world's best tripel, we won't be telling the truth.'

THE SONG

Lady Gaga, 'Bad Romance', Interscope Records, 2009, *Electropop/Experimental Pop*

It's quite a thrill to include this song here, especially as a pairing with such an outstanding, iconic beer. It may feel like Lady Gaga isn't very beery, and that the audience for her music is not the audience for Belgian Trappist beers. I feel this is appropriate, because this kind of dissonance is what Gaga is all about.

Like all true artists, Lady Gaga is constantly absorbing influences, always questing and learning. She wrote this while touring Scandinavia and Central Europe. Musically, she was intrigued by German house and techno, and wanted to incorporate these elements into her evolving sound. Lyrically, the opening line, 'I want your ugly, I want your disease,' establishes that 'Bad Romance' is a dark mirror to the way conventional pop songs invariably go on about love and sexual desire. It's about wanting to have sex with a close platonic friend, and ruining that relationship. It's about her need for love from her fans. And it's about the fears and insecurities, the 'monsters' that plague her in the strange word of the tour bus. To wrap all this up in what the American Psychological Association determined to be officially the catchiest earworm in the world is quite a feat.

THE PAIRING

Having done this for fourteen years, and having hopefully proved that I'm not just making it up, sometimes the match comes first, then I have to go back over it and figure out why it feels right. This is not a pairing I've ever used at an event, but the song is a classic, even if it's wasn't previously in my own library. And so is the beer.

Apparently 'a lot of whisky' was involved in the song's composition, but I'm still claiming it as a beer song. To substitute for whisky, it would have to be a beer as complex as this one, and I realised that's where the match came from. Both the beer and the song are strong, sophisticated and complex. There's just so much going on in both, so many layers. You have to keep going back to them, getting new things every time.

Rudgate Ruby Mild with Young Fathers 'Rice'

THE BEER

Rudgate, Ruby Mild (4.4%), *Mild*, UK

I once write a piece for the *Sunday Times Magazine* where I made a joke about no one drinking mild any more. I now know that at least thirty people do drink mild still, because every single one of them wrote to tell me how wrong I was.

Mild just doesn't seem to be able to catch on as a style these days, and I don't really understand why. I haven't just included this example here in an attempt to clear my name: it's an amazing beer and should be far more famous than it already is. It was named CAMRA's overall Champion Beer of Britain in 2009, 2012 and 2017, and was runner-up in 2008, 2011 and 2015. Cynics dismissed this as CAMRA merely trying to revive interest in mild as a style. As is the case every year, the most vociferous critics hadn't actually tasted the beer.

The aromas drag you in. There's rich red fruit, caramel, roasted grains, chocolate, even a hint of smoke. If you were trying to pair this in terms of musical pitch, it runs the whole length from low to high. It sits mainly lower and darker on the palate, with toffee, caramel and nuttiness, but there are some high-pitched fruity notes too, all wrapped up in a smooth, velvety finish.

I wonder if it's the name 'mild' that puts people off? Perhaps if it was called 'chewy fudge monster' it might be a really cool craft beer? Maybe Rudgate should do their own scientific experiments about possible correlations between naming and design, and perceived flavour and appeal.

Young Fathers, 'Rice', Ninja Tune, 2023, *Electronic/Pop/R&B*

I swear I'm not relying on Wikipedia to get the blurbs for this book. But I do occasionally check it out for some details such as record labels and release dates. It attempts to define the genre a band plays in, and the entry for Young Fathers is hilarious. It describes them as neatly fitting into 'Lo-fi, indie, soul, alternative hip hop, indietronica [a new one on me], art pop, avant pop' and 'noise pop'. The incredible thing is, I can think of at least one or two genres they've missed there. A complicated-step-relative-sort-of-cousin of Liz's went to music college in Edinburgh at the same time as Young Fathers, and she reports that they were more of a boy band back then. I saw them live a couple of years ago – one of the best gigs I've ever been to. And under the primal, swampy excitement, hallucinogenic rhythms and general sense of a revolution being incited, those earnest, sweet harmonies occasionally remained intact.

Each new album from them gets denser and richer. There's so much going on in this track, I imagine it as the kind of music you might hear if you were to find yourself in a shebeen on an asteroid mining colony in the year 2345. 'Rice' is the opening track from their fourth album. It might as well be the intro music to the film ten years from now that redefines what a global, kinetic, breathless action movie can be. It probably will be.

THE PAIRING

This is about density, complexity and layers. It's the kind of beer where I can get so lost in the nosing of it, I almost forget to drink. The song's density means, similarly, that you can't get everything on one go. There's a lot going on in the pairing of low musical pitch and the corresponding flavours of chocolate and smoke, but it goes beyond that, bleeding out of the edges. Instinctively, it's a contemplative pairing that would seem to suit an armchair and an open fire. But it's also dragging you up to dance. How can it do all that at the same time? Like I said, there's an awful lot going on.

Orval with Patti Smith 'Land'

THE BEER

Orval (6.2%), *Belgian Blonde*, Belgium

There are few, if any, beers that are more revered than Orval. Even within the pantheon of Belgian Trappist ales – a sort of Champion's League fantasy team of the beer world – if we're talking simply about the beer itself, rather than bragging rights and enforced scarcity, it's probably the favourite of serious beer fans worldwide.

We think of Trappist brewing as having roots that are centuries old, and in a way they are, but the beers we know now are much more modern creations. The current brewery at Orval dates back to 1931, and Orval was first shipped a year later.

Orval is fermented once using ale yeast, and then again in conditioning tanks with a hint of *Brettanomyces bruxellensis*, the wild yeast found in the Senne region on the outskirts of Brussels. It's dry-hopped, and then re-ferments once again after it's been bottled. It's a long process, and the beer continues to evolve once it leaves the brewery. Some bars sell old, 'out of date' Orval for more than the fresh stuff.

The resulting taste is incredibly complex, without being overwhelming or difficult. It's dry. It's fruity. It's bitter. It's spicy. It's both vinous and earthy, funky and yet somehow clean and integrated. Orval is one of those remarkable beers that truly does reward careful contemplation. There's something new to find in every sip.

THE SONG

Patti Smith, 'Land', Arista, 1975, *Punk Rock*

Patti Smith can seemingly do anything, but her centre of gravity is probably poetry. In the late 1960s, frustrated by the barriers on performance poetry, she recruited guitarist Lenny Kaye to push past those barriers. Poems gradually turned into songs, but in a very loose-limbed, fluid way, evolving with each performance.

The simple three-chord riff of 'Land' is taken from 'Gloria', recorded by Van Morrison's band Them in 1964. (The album *Horses*, on which 'Land' appears, also contains a more direct cover of the same song.) But the heart of the song is also a cover version of 'Land of 1000 Dances'. First written by Chris Kenner in 1962 and subsequently covered by various artists, the song celebrates sixties dance crazes such as the Twist, the Alligator, the Mashed Potato, the Watusi and the Pony over a similar three-chord riff to 'Gloria'.

This sort-of-cover of 'Land of 1000 Dances' was apparently used as an audition for potential new band members. They'd start playing the riff, and Smith would go wherever the energy took her, combining fragments of the original lyrics with the story of Johnny, according to Smith, a 'pre-punk rock kid … entering the world, ready to take it on … a metaphor for the birth of rock 'n' roll.' The song – it's more like an incantation or spell, really – would sometimes last for twenty minutes, if the auditioning musician didn't collapse first. The album version is a mere 9'25", so Smith's vocals are multi-tracked, different ideas speaking over each other, to fit them all in.

THE PAIRING

I first put this match together in a very instinctive way. They just felt related. Both are, to me, about wildness. If Patti Smith truly were casting spells with her vocals, then Orval's distinctive chalice glass looks like the kind of thing that might hold a magic potion. Orval's use of Brettanomyces yeast is skilful: the wildness is there, but it's tethered, threatening to break free even though it never does, and that's how I feel listening to 'Land'. Additionally, the fruitiness present in Orval feels like it matches the pitch of 'Land', while the mild dissonance in the song fits the faint sour note in the beer.

More than that, though, I like the idea that both are constantly evolving. I've only ever seen Patti Smith live once (in a tiny room in Dylan Thomas's boathouse in West Wales, as part of an audience of twenty, one of the greatest gigs of my life) but I imagine no two performances of 'Land' are ever quite the same. Both the beer and the song are like living creatures, never quite still.

Abbaye de Rochefort 10 with Johnny Cash 'Hurt'

THE BEER

Abbaye de Rochefort 10 (11.3%), *Quadrupel,* Belgium

Every time I drink this beer, it feels like a special occasion. I feel like I should dress smartly and comb my hair before I open it. It's a beer that commands respect, that deserves to be addressed as 'sir' or 'madam' (I suspect the latter.) The main beer styles brewed by Trappists and other abbey brand names are dubbel and tripel. Inevitably then, the next one up has become known as a quadrupel.

While the abbey at Rochefort has roots as far back as 1595, Napoleon's occupation of Belgium closed its Trappist monasteries. Rochefort was rebuilt in 1887, and the current brewery was opened in 1899.

But this one feels like a timeless beer, a perfection of centuries of art and dedication. If you've never encountered a beer like this before, it will change your entire worldview of what beer can be. You lose yourself in nosing it. Each sniff reveals more. A lot of chocolate and creamy coffee, a bit of fruit. Is that a hint of wood? There's spice, citrus and chocolate on the palate, a rich, silky, warming body. If you're a non-smoker, I imagine this is the closest you'll get to enjoying the experience of smoking a Cuban cigar.

THE SONG

Johnny Cash, 'Hurt', American Recordings/ Universal, 2002, *Folk Rock*

Johnny Cash's late career renaissance has been much-documented and widely lauded. This was its highlight. Cash's voice was always rugged. Here it's ruined. What was always hard has now ossified completely, his mouth and lungs turned to

teak. Much was made of Cash's age –
he was seventy-one at the time – but
that voice sounds older. It's the voice
of a life lived hard, simultaneously
vulnerable and strong.

This track is a cover of a song
by Nine Inch Nails. The lyrics are
unflinching. With themes of self-
harm, drug-taking, loss and betrayal,
it's not a song I'll ever use to lighten
the mood at one of my pairing events.
But that voice, and the knackered
piano that tolls like a cracked bell in
a Wild West graveyard, give the song
power and dignity even as Cash sings about the bleakest sense of loss.

He was originally reluctant to sing it, thinking it wasn't his style. Similarly,
songwriter Trent Reznor felt the song was so personal to him, hearing Cash sing
it would feel like 'watching my girlfriend fuck somebody else'. But when Reznor
heard the finished result, he described 'Tears welling, silence, goose-bumps …
Wow. I just lost my girlfriend, because that song isn't mine anymore.'

THE PAIRING

Some of this is about context and intuition. Both the song and the beer are very
serious. Both can be described as very deep and heavy. The video and the album
sleeve were in black and white: the beer is so dark brown it's almost black.
We feel like we're at the bottom of some dark pool here.

But there's also a crossmodal match based on bitterness. The beer is strong.
And while bitterness isn't its defining characteristic, there's coffee grounds and
dark, bitter chocolate in there. In the crossmodal mapping of flavour to musical
styles, these pair precisely with that voice, and the tonal pitch of the song in
general.

If this all feels a bit depressing, the experience of drinking this beer is
anything but. It's joyous. It makes you feel special. And I'd argue the bleak song
does not make you feel bleak in turn. It elevates you. You're listening to something
really good, really powerful. Sit back and let the pairing wash over you.

Acorn Barnsley Bitter with Grimethorpe Colliery Band 'Gymnopédie No.1'

THE BEER

Acorn Barnsley Bitter (3.8%), *Bitter*, UK

When does a single beer become a beer style? It happened to Pilsner lager, and more recently to ESB – beers that were so ground-breaking, so popular, that they were copied by brewers around the world. Coming from Barnsley, I'm hopelessly biased, but I'd argue that Barnsley Bitter is just about there too.

You can't trademark the name of a town, and there are at least two Barnsley bitters being brewed in Yorkshire today, both to a similar style, both excellent. There used to be more.

Acorn's take is the most widely known, and the most award-winning. It's light and sessionable, as you'd expect from a 3.8% bitter. But the characteristic that sets Barnsley bitter apart is a chewier, fruitier body. There's chocolate in the nose and red berry fruit on the palate. There's also, unmistakably, parkin. A traditional Yorkshire ginger cake, parkin was a staple of bonfire night when I was growing up in Barnsley. It's made of flour and eggs, butter, golden syrup, treacle and love and warmth. Barnsley bitter gives you fleeting hints of parkin as it goes down, and has the advantage of being liquid, so there's no danger of it costing you any loose teeth, like there is with the cake. This is a very wholesome beer. One that will keep you going, especially when it's cold and wet.

THE SONG

Grimethorpe Colliery Band, 'Gymnopédie No.1', Parlophone, UK, 1989, *Brass/Ambient*

Brass bands were the music of my childhood. You simply heard them everywhere. There was a lot of pride in the fact that burly men who didn't say much and hid their emotions could produce work of such beauty.

Like Bavarian 'oompah' bands or Scottish bagpipes, there was always a certain amount of ridicule about the genre, brewed of class snobbery, and the fact that, yes, sometimes you heard something that wasn't very good.

But the best were brilliant. Most, if not all, of the collieries around Barnsley had their own bands. But Grimethorpe were the best. They were the real band who played in the 1996 movie *Brassed Off*, their home village fictionalised as 'Grimley'.

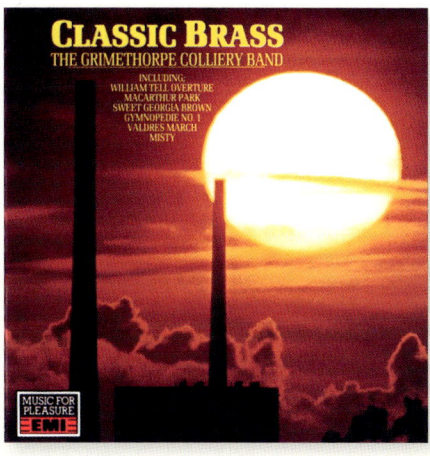

Brass was always entertaining for the span of genres the bands attempted to cover. They were incredibly versatile. But I was surprised and delighted, when choosing a song for this book, to find that they'd had a go at Eric Satie's 'Gymnopédie No.1'.

With his three Gymnopédies, French composer Satie essentially invented ambient music. There are stories of him playing them live, and becoming distressed that people were politely paying attention to him. He wanted them to carry on chatting, to be their background.

I wanted to use Satie in this book. And I wanted to use an example of brass band music. That I can do both in one go shows just how much brass band music can be misunderstood and under-appreciated.

THE PAIRING

In crossmodal pairing experiments, brass is associated with bitterness, mainly because it's seen as less pleasant than the sweet piano. The flaw in the experiments is that we're used to hearing a piano on its own, but not a brass instrument. So the brass isn't going to sound right when used on its own. For me, this tune proves that brass – like the bitterness in the beer – can be soft and mellow.

But the main reason for this pairing is context. The music *smells* of Yorkshire for me. The beer tastes of Barnsley. As I explored in my book *Clubland*, you have a very complicated relationship in adulthood with a home town you couldn't wait to leave when you were growing up because you didn't fit in. You can never go back, but you carry it with you. This pairing is about growing up in Barnsley, and the hidden beauty peeking out from beneath the soot.

Cantillon Rosé de Gambrinus with 808 State 'Cubik'

THE BEER

Cantillon, Rosé de Gambrinus (5%), *Lambic*, Belgium

It took me a while to get to grip with lambic beers. The first few times, I rejected them completely. Then I thought: if I want to write about beer full-time, I need to come to terms with them. It's worth persevering. The first time you try a lambic, it pulls your face inside out and fractures your sense of an objective reality. The second sip starts to make sense, even though you're crying by now and calling for your mother. The third reveals beer's equivalent to Champagne.

Lambic is one thing. Rosé de Gambrinus is another. At the risk of introducing too many analogies with other drinks, I once heard it described as the Laphroaig of beer. It's a perfect comparison: I love the medicinal, TCP hit of Laphroaig, the attack of its intense peatiness. There are many malt whisky aficionados who feel it is just too much, crude and inelegant. It's a valid opinion. But by the time I'm drinking Laphroaig, I'm often crude and inelegant myself.

Rosé de Gambrinus is a blend of different lambics along with 200g of raspberries per litre. Hilariously, the brewery describes it as 'slightly' acidic, which is like saying Elon Musk is 'slightly' weird. Rosé de Gambrinus is an intense hit, even by the standards of so-called 'sour' beers. Lambics themselves are often not as sour as we make out: they're drier and mustier. My theory here is that the raspberry juice, which has all the fruit sugars fermented out to dryness, adds to the intensity of the acidity.

If you really can't face it as a drink on its own, it makes an awesome garnish for oysters if you finely chop some shallots into it.

808 State, 'Cubik', ZTT Records, 1989, *House*

From the opening guitar squall (at least, I think it's a guitar, but it might be a duck being garrotted) 'Cubik' is an attack. Each note is followed by another note that you wouldn't expect, an anti-melody that is still somehow fluent. It stabs you in the ear and shouts in your face, and then does something completely random that shouldn't quite fit, but does. It's a cross between a chaotic house party after midnight, the bit where people let off fire extinguishers and smash things, and those YouTube videos

cubik

where someone puts a brick inside a washing machine. And yet, the main stabby, squelchy hook is like when you have an itch in the middle of your back that you can't quite reach, and then someone scratches it for you, only it's in your brain. At the time, 808 State were gaining a reputation as one of the leading exponents of ambient house. 'Cubik' is the sound of them refusing to be put into a square box.

THE PAIRING

This is textbook pairing, even if it's one that might have made the original academic research subjects cry. The work of Bruno Mesz and others established that sourness/acidity pairs with music that is high in pitch, and highly dissonant. When I've attempted to make pairings between 'sour' music and sour beer, something interesting happens. It's important to match the *degree* of sourness with the right degree of dissonance. Play a song like this with a beer that has gentle sourness, and that gentle sourness completely disappears. If the degree of dissonance is itself dissonant, the pairing doesn't work.

So this is a pairing in which the sourness of the beer and the dissonance of the music are both quite assertive, both in yer face, (which, curiously enough, is the title of another 808 State song.) Both the beer and the song excite me, even as they slightly scare me. A beer from a very traditional, sleepy little Belgian brewery, and a band from the heart of the late eighties/early nineties Madchester scene, are doing exactly the same thing as each other.

Fullers ESB with Lorde 'Tennis Court'

THE BEER

Fullers, ESB (5.6%) *Extra Special Bitter*, UK

ESB stands for 'Extra Special Bitter'. It was first brewed in 1971, at a time when many brewers had an 'ordinary' and a 'special' bitter. Fuller's already has a special in London Pride, so when they wanted to amp it up, it had to be extra-special.

ESB has won many awards since. But perhaps more revealing of its nature is that it's been imitated so many times globally, Fuller's eventually gave up trying to protect their ownership of the name. ESB, which began as the name of a single beer, is now the name of a global beer style of which Fuller's ESB is the platonic ideal.

I write about this beer whenever I have the chance, mainly because it challenges our ideas of how we should write about flavour. We often describe beers as malty or hoppy, but we don't talk enough about the interplay between the two, the totality of the ingredients in perfect balance. To talk about this beer in terms of malt or hops would be to miss the point of it. I could say it has hints of caramel, or it echoes red wine, or that it has a grassy note. But to reduce a beer like this to its component parts would feel like the behaviour of a psychopathic child who performs a live dissection on his pet rabbit to see how it works, and gets upset when it doesn't want to play with him anymore.

THE SONG

Lorde, 'Tennis Court', Universal, 2013, *Alternative Pop*

Ella Marija Lani Yelich-O'Connor was born in 1996, in Aukland, New Zealand. By the age of six, she was recognised as a prodigy. By the time she was twelve, she had a record deal. Working as 'Lorde', in collaboration with producer Joel Little, she recorded an EP featuring the song 'Royals'. The track was released as a single in 2013, when she was sixteen, and sold ten million copies, becoming one of the best-selling singles of all time.

Lorde   Tennis Court EP

'Tennis Court' came from Lorde's second EP, and went on to be the opening track on her debut album, *Pure Heroine*. It's a remarkable song. Musically it's stripped back to sparse beats and growling synth chords. Lorde learned to write lyrics from paying attention to how short fiction works, and the story she tells here, after the global success of 'Royals' at such a young age, is of how her life is changing. She's just about to get on her first ever plane. She knows a new life of stardom awaits her, and she knows what the personal cost of that will be.

It's an incredible and impossible mix of self-conscious naiveté and a world-weary knowingness. From a sixteen-year-old. And it's catchy as hell.

THE PAIRING

The beer match came to me in a flash while I was listening to the song, and I had to work back from the instinct to figure out why. It's about the relationship between flavour and pitch. The song simmers along in a low pitch somewhere around where flavours of chocolate and dark fruit have been found to gather. There's also a proven association between low pitch and bitterness. The bitterness in ESB is gentle, but it's there, and it works, especially around the slowed-down sampled 'yeah's that punctuate the song. Contextually you might not think an artist like Lorde and a traditional bitter such as ESB go together. But the science suggests they do. Turns out the science is right.

Mackeson Stout with The Shadows 'Wonderful Land'

THE BEER

Mackeson (2.8%), *Stout*, UK

Mackeson's brewery in Hythe, Kent, released this stout in 1909 to celebrate 240 years of continuous brewing. Ten years later, it was acquired by Whitbread. By the 1950s, Mackeson accounted for half of Whitbread's considerable volume.

Mackeson is a milk stout, brewed with lactose. This can't be fermented by brewing yeast, so the beer remains sweeter and has a thicker mouthfeel – ideal if you want a lower strength beer without feeling like something is missing. It used to be recommended for nursing mothers. With the arrival of TV advertising, Mackeson prospered with the line, 'Looks good, tastes good and, by golly, it does you good.' New rules around claiming health benefits for beer eventually killed that.

When I used to work with Whitbread, they had a class of beers they referred to internally as 'dogshit brands', which were beers they owned but were not interested in. Whitbread is now part of Anheuser-Busch InBev, and they obviously believe Mackeson belongs in this category. It's had no promotion or marketing for decades. It isn't mentioned once on their website, and doesn't even feature in a photo of their entire UK beer range. And yet, it still clings to life, lurking at the back of supermarket shelves, enjoying a cult following.

THE SONG

The Shadows, 'Wonderful Land', Columbia, 1962, *Instrumental Pop*

The Shadows were huge. If it hadn't been for the Beatles, more of us who are under seventy would probably remember just how huge they were, before the world changed.

Okay, so they started off as Cliff Richard's backing band. But before he got weird, Cliff was Britain's first proper homegrown rock 'n' roll star. After he was hived off to become a solo pop idol, the Shadows were free to carve their own furrow as an innovative, electric guitar-led instrumental band.

How huge were they? Well, 'Wonderful Land', a two-minute-long instrumental performed by four blokes who did a cheesy synchronised dance, fronted by Hank Marvin – a man who had the name of a rock star but looked like an accountant behind his thick-rimmed glasses – was number one in the charts for eight weeks, at a time when you had to shift millions of units to make that happen. Only Elvis Presley and the Archies managed to equal that in the sixties. Nope – not even the Beatles. The title refers to America. The soaring guitar line sings of an idealised dream of it.

THE PAIRING

This one's for my dad. He didn't drink much beer, but when he did, it was Mackeson. He loved music. But like so many people, his exploration of it pretty much arrested not long after his teens.

There were one or two exceptions. When I started buying my own music, he heard 'Blue Monday' by New Order and said, 'What's this? Alright, innit?' It was. He came into my bedroom once to ask what album I was playing. 'It's called "Treasure", by the Cocteau Twins,' I said. He nodded and left. Afterwards, whenever he was driving me back up to St Andrews University at the start of term (he insisted), we'd get onto the A1, and he'd say, 'Put t'Rocker Twins on.'

But with the honourable exception of Jean Michel Jarre, his own preferences were early British rock 'n' roll, Shirley Bassey, and James Last's orchestral easy listening. His favourite band were The Shadows. Dad would play up his thick Yorkshire accent when he was happy. Famously, we flatten our vowels when speaking full broad Yorkshire. And dad would growl the levelled 'a' in 'Māckeson' and 'Hānk' in exactly the same way, from the back of his throat. When The Shadows came on telly, there'd be one guttural grunt of 'Hānk', almost like a clearing of the throat, and then he'd sit in silence, his eyes moistening.

And you know what? I'll be damned if there isn't a crossmodal match here too. The sweetness of the stout pairs perfectly with Hank's soaring guitar parts. Cheers, Dad.

Petrus Red with Claudio Monteverdi 'L'incoronazione di Poppea'

THE BEER

Petrus, Red (8.5%), *Fruit Beer*, Belgium

Judging beer in competition is not as much fun as it sounds. Most beers are indifferent. The logistics of a global competition mean they're quite old and have been knocked about a bit by the time you taste them. There are some that are truly awful. When you get one of these, you have to taste it as much as the great ones and describe in detail exactly *why* it's horrible. And then, every now and again, you're judging a style you don't care for and find an absolute gem. You like it so much, you have make a note of its code number so you can ask what it is when the judging forms have been handed in.

Petrus is a traditional Belgian brewery that has observed craft beer embrace 'sour' as a style, and rebranded, saying, 'Hey, we were doing this first! Look how cool we are!' It grates with me. Because as I've said elsewhere, the 'sour' label does any decent beer in this style a serious injustice. Petrus Red is a blend of pale ale aged in oak foeders for two years, and dubbel beer blended with cherries. Yes, it has some sourness (it scores 2/5 on the 'sourness scale' on the brewery website) but it also has some sweetness from both the fruit and the dark, chocolatey dubbel. It's complex yet refreshing, sour and sweet – a perfect example of how skilful blending can create a beer that's greater than the sum of its parts.

THE SONG

Claudio Monteverdi, 'L'incoronazione di Poppea, SV308, Act 3: Pur ti miro,' 1643, *Opera*

I said on Side One that I didn't like opera. Here's the exception. Like Vaughan Williams's 'The Lark Ascending', this is a piece you can appreciate and understand without having a detailed knowledge of the backstory or conventions of the form.

'L'incoronazione di Poppea' (the Coronation of Poppea) was the last opera that Claudio Monteverdi wrote, completed just a few months before his death in 1643. It was one of the first operas based on historical events, telling the story of how Poppea Sabina, a mistress to the Roman Emperor Nero, was able to marry him and become crowned Empress of Rome. Historically, she has often been portrayed (by male historians, obviously) as scheming and devious. How else could a woman achieve such success?

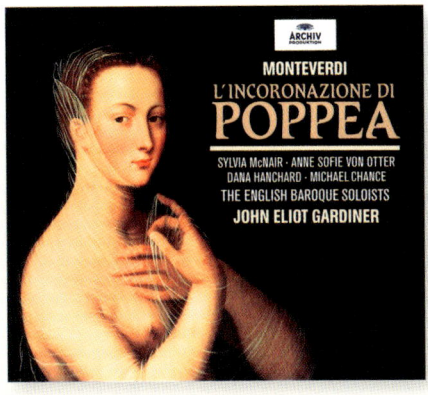

I've taken a deliberate decision not to find out where this bit comes in the opera, or what it is supposed to depict. I almost feel bad for finding out as much about it as I have. It's not about that. It's about the crisp, autumn mornings in the house in Norfolk which Liz and I moved to in December 2023. It's on a playlist she puts on when she's doing something like baking gingerbread or doing the deep prep for dinner when it's her turn to do Date Night. It's about the way these female voices sound as they echo in our cold kitchen, which has had nothing done to it since the 1970s, and how I like to join her there at the weird, old kitchen table as the sun comes round far enough to shine in through the window. If you don't have comparable moments – either alone or with a partner – I urge you to create some to go with this tune.

THE PAIRING

This one is straight down the line sweetness. Sweet maps with music that is high-pitched and melodic. The voices here (English baroque soloists Sylvia McNair and Dana Hanchard) are a slam dunk for that. There's a little bit of sourness in their harmonies too. I like to elaborate on the basic theory of what sweetness corresponds with in music. Close vocal harmonies are high-pitched and melodic, and therefore I automatically count then as sweet. Experiments in crossmodality talk about 'hedonic transference' – we like sweet flavours, so we like music that corresponds with them. I feel something beautifully bittersweet in the way these two voices wrap themselves around each other, playfully and yet wistfully, formal yet fresh-sounding.

Heritage Brewery Masterpiece with This Mortal Coil 'Song to the Siren'

THE BEER

Heritage Brewery, Masterpiece (5.4%), *IPA*, UK

As a beer writer, the question I'm asked more than any other is, 'What's your favourite beer?' No writer I know can answer it. I'd suggest we all have a top ten, maybe even a top five. The odd beer will come and go over time.

For almost twenty years, Worthington White Shield was in my top five. Maybe even my top three. Then, in 2023, owners Molson Coors axed it. They didn't want to get rid of it from their portfolio: they just wanted to stop brewing it. They didn't want it, but no one else could have it either.

So White Shield doesn't exist any more. But Masterpiece does. Masterpiece just happened to be formulated by Steve Wellington, the brewer who knows more about White Shield than any other person on earth. This is a very, very similar beer. One could almost say it's the same recipe under a different name, brewed with a slightly different yeast.

It's dark and malty, a little spicy, with hints of orange marmalade, constantly evolving on the palate and developing new layers. It reminds me very strongly of another beer I used to know.

THE SONG

This Mortal Coil, 'Song to the Siren', 4AD, 1983, *Ambient*

This is a song that changed my life completely as soon as I first heard it. It literally stopped me in my tracks. I couldn't move, hardly dared breathe, until I was sure it had finished. It's simply the most beautiful thing I have ever heard in my life. This Mortal Coil was a project whereby Ivo Watts-Russell, head of independent record label 4AD, got his roster of artists to cover some of his favourite songs, with a consistent production aesthetic that made his diverse tastes sound like they were all in the same genre.

Elizabeth Fraser and Robin Guthrie, otherwise known as the Cocteau Twins, covered Tim Buckley's 'Song to the Siren'. Buckley's original is very good indeed – a hippyish, folky love song built around the analogy of a sailor being lured onto the rocks by the mythical siren. You can tell a song is this good when it's been covered by everyone from Bryan Ferry, Robert Plant, Sinead O'Connor, George Michael, Wolf Alice and Garbage to Alfie Boe and Half Man Half Biscuit.

But it was This Mortal Coil's version that rescued the song from obscurity. It's so sparse, it's hardly there. Guthrie's guitar sketches a few quick pencil lines to outline the bare bones of an atmosphere. Fraser's voice, backed up by some synth-produced echoes, does what only her voice can do. Did anyone involved think that she was, literally, the siren in the song? It would probably be too clichéd to admit it. But she enslaves me and enfolds me every time I hear it.

THE PAIRING

You've done me the kindness of reading this far, so I'll be honest with you. This is my book, my rules. I've been fighting to get a deal on this book for nearly fifteen years. Now it's finally happening, there is simply no way this manuscript is leaving my desk without this song and this beer in it.

Do they follow crossmodal principles or have some kind of contextual or linguistic pairing? Who cares? This is a song I want played at my funeral. And this is the beer I want people to be drinking while they're hearing it.

So have I just debunked all the science I've been trying build up? Not at all. This is an example of what I would consider a confounding factor in a strictly scientific context: if you have a previous relationship with a beer or a piece of music, it affects your perception of the experiment. Here, my intense passion for this beer and this song mean I simply cannot be objective about their pairing potential: the liking carries over. It's called hedonic transference. That sounds scientific enough for me.

Titanic Plum Porter with Astrud Gilberto 'I Haven't Got Anything Better To Do'

THE BEER

Titanic, Plum Porter (4.9%), *Porter*, UK

When Titanic first brewed Plum Porter in 2011, it was supposed to be a one-off. It now accounts for more than half the brewery's total beer sales.

A few years ago I remember talking to Titanic Managing Director Keith Bott, and he seemed a bit bemused by it. He thought it was a bit of a novelty, adding fruit to beer, and couldn't understand why everyone liked it so much.

I suspect he's changed his tune. Titanic now also do a cherry porter, a cappuccino stout, and a chocolate and vanilla stout, as well as very special, limited editions of plum and cherry porter which are punchier as well as being more refined.

The original plum porter works really well because of flavour connections. In a straight porter, you often find hints of dark forest fruits. Why not tease out and develop those hints with the fruit itself? In many cases, the answer to that question is that hints are enticing, being hit with an unsubtle fruity sledgehammer is not. Nuance can be nice. So it all comes down to the skill of the brewer as to whether you get a fusion of beer and fruit that's harmonious or some kind of nasty fruit purée having an identity crisis.

Titanic Plum Porter is, obviously, the former. If this beer were a house, it's like it's had an extension added on. Something new, but in keeping with the original style.

Astrud Gilberto, 'I Haven't Got Anything Better To Do', Verve, 1969, *Baroque Pop*

If this song had been written in the age of social media, it would have been called 'I'm not crying. You're crying'. It has that same sentiment of hot embarrassment that someone can see you're upset, making you revert to childish logic.

Happily, the song was written in 1969. The emotions are the same. The childish logic is still there. But it presents as infinitely more sophisticated and stylish.

The intro music evokes a sense that it's leading you through the door of some quiet Parisian café, past empty tables to a corner in the rear, where Astrud sits alone. She tells you her story between smoking Gitanes and angrily wiping away tears with the palm of her hand. She reveals how broken-hearted she is, and why, by denying accusations that you never made. It was nothing special. Nothing at all. She doesn't care about him. Never thinks about what he's up to or who he's with. Never remembers their time together. Except when she has nothing better to do. Which is, of course, always. With that voice, and the strings that support it and comfort it – a broken heart has never sounded so cool.

THE PAIRING

I put this pairing together for the book while I was looking to diversify the styles of music it covers. Then, a couple of weeks before writing this, I was doing an event at a beer festival where Plum Porter was pouring, so I decided to try it out on an audience. I think it helped that it was Valentine's Day.

It's clearly all about sweetness – the fruit, and the instrumentation. But it's also tinged with bitterness – the sentiment of the song, the dark, rich flavours of the beer. And there's something textural there too – the silky smoothness of her voice and the almost vinous mouthfeel of the beer.

But what was remarkable in the event was that the song massively brought out the plum character. It was there. Then, during the song, it came to the front to play a solo. After the song, it fell back into line. It was quite something. A hell of an opener.

Salopian Lemon Dream with Skeeter Davis 'The End of the World'

THE BEER

Salopian, Lemon Dream (4.5%), *Golden Ale*, UK

Salopian are one of those breweries I always turn to when I'm looking for a paragon of a particular beer style. If they've decided to make one, it's going to be good. They're not showy or flashy. They just brew great beer and win loads of awards with it. Since 1995, they've grown steadily, evolved steadily, kept up with current trends in brewing without ever seeming to jump on a bandwagon, and quietly become one of the most admired breweries in the UK.

It's telling that Lemon Dream is part of their core range of six, and is brewed with the addition of organic lemon juice. Like the other fruit beers that have made it into the selection for this book, the lemons here take a characteristic that was already in the beer and gently draw it out and extend it. First brewed in 2001, it was a prediction of where the craft beer market would go a little over a decade later.

The hops are slightly different in the cask version from that which goes unfiltered into kegs and cans, but it's still pretty much the same beer. Cleverly then, it's a beer that fits fair and square into both the quirkier end of traditional British cask, and modern US-inspired craft.

THE SONG

Skeeter Davis, 'The End of the World', RCA Victor, 1962, *Country*

Skeeter Davis was one of the first solo female performers to become a country music star. By doing so, she became a profound influence on future legends such as Tammy Wynette and Dolly Parton.

Like them, as well as nailing country, she also crossed over into pop, and 'The End of the World' was her biggest hit, later covered by acts as diverse as The Carpenters and Patti Smith.

That's because, while it's a country song, it's much more besides. In the chorus, I hear the kind of melody that would define the ballads of every girlband and boyband from the Ronettes and the Osmonds through to the twenty-first

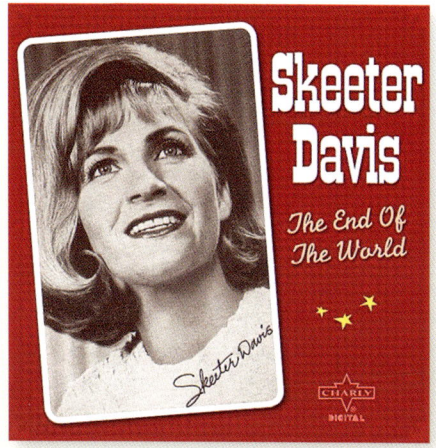

century. I'm also fascinated by a perfect example of that weird convention where, in the middle of the song, the words are said rather than sung, as if this gives them some deeper profundity. At the time, it worked. Now I find it hilarious.

These elements make the song quite kitsch. As well as country and pop, it belongs in a nostalgic pantheon of songs that feel firmly rooted in their era. This made it perfect for inclusion in the *Fallout* video game series, where a post-apocalyptic wasteland is full of references to the future as it was imagined in 1950s America. This, and other songs in the same vein, can be experienced as both funny and poignant at the same time.

THE PAIRING

The strongest and most effective pairing ideas have cycled around time and again in this book. So here's one final example of how music can correlate and enhance both sweetness and high pitch.

It provides a useful look at the difference between sour – which is often associated with citrus – and the character of citrus itself. Get a squirt of pure lemon juice on your tongue, and you might describe it as sour, but that's imprecise. It's sharp, yes. And acidic. But as I keep saying, sour is not an adequate description. Eat fresh citrus, or drink lemonade, with enough sugar to temper it, and there's a meeting between sweetness and acidity.

Sweetness is also high-pitched. This pairing between a lemon-accented beer and a high-pitched, mellow and sweet voice explores the fruity territory where sweet and sour meet and cross over. The musical pairing is helped by the sweetness in Davis's voice and the melody gaining an extra layer of slight dissonance from the kitsch that's developed around it.

Mantle Dis-mantle with John Prine 'Summer's End'

THE BEER

Mantle, Dis-mantle (5.8%), *Strong Amber Ale*, UK

When I do my big beer and music show at the Green Man Festival, my challenge is to work with the bands on the line-up and the local beers in the cask ale tent. The beer line-up has evolved significantly over the last decade, but some beers are there every year. I've used this one more often than any other.

Mantle are a small brewery based in Cardigan, West Wales. Most of their range is similar to that of other small breweries – pale ales, and best bitters, a lager-type ale, and so on. Dis-mantle is their showstopper.

It's one of those delightful beers that bucks all trends and doesn't quite fit in to a contemporary map of beer styles. At 5.8%, it has a deep body and thickness to it, but at the same time it's crisp and refreshing. It combines vibrant, grassy, spicy hop aromas with the deep caramel of a best bitter and the chocolatey hints of a porter or stout. It's kind of like an ESB, but deeper. It's unique. And because of this, it can pair with anything from modern folk to heavy metal. It's become like an old friend to me, with an excellent taste in music, always waiting to greet me at every festival. It's the beer I end the night with – if there's any left – as the final act on the main stage lights up the Brecon Beacons.

THE SONG

John Prine, 'Summer's End', Oh Boy Records, 2018, *Country*

One of my favourite things about music is the archaeological digs it sometimes takes us on. I love the music of Brian Eno. Spotify knows this, and the algorithm directs me to any new Eno releases. Two years ago, it flagged up an album by Fred Again.., who I hadn't heard before, in collaboration with the master. There's a snatch of a song buried under layers of samples and distortion, a whispered plea to 'come on home'. It returns on the album's final song, a spectral song of loss and longing. Fred Again.. works with layers – the principle of building and layering is more important than any individual instrument. The final track, 'Come on Home', is a cover – or maybe a sample? – of Adrianne Lenker, lead singer and guitarist of the band Big Thief, doing an acoustic cover version of a song called 'Summer's End', by John Prine, in the midst of the Covid-19 pandemic. She recorded it while Prine was critically ill in hospital with Covid at the time. She sang it as a kind of vigil to him, clearly upset during the performance, which can be found on YouTube. Prine died shortly after.

I hadn't heard of John Prine until this twisting, turning tunnel led me to him. I'm glad it did. He was a musician's musician, adored by people like Tom Petty, Bruce Springsteen, Bob Dylan, and Kris Kristofferson, who said that Prine wrote songs so good that 'we'll have to break his thumbs'. 'Summer's End' is mournful and bittersweet, deceptively simple. A song of layers. The sense of loss in the song is mirrored by the loss of an artist I never knew until he was no longer here.

THE PAIRING

This is all about low pitch: the deep, malty notes, halfway between a best bitter and a porter, and a voice that's careworn, that sounds like the lines in an older person's face. The pace of the song adds to the slow, contemplative mood, which suits a beer that can't be rushed. And there's a deeper synaesthetic match, too. If you accept that certain music sounds and vocal stylings can be called 'The Blues', then this for me is 'The Browns' – that voice to me sounds dark brown, the same colour as the beer.

Thornbridge Jaipur with New Order 'Temptation'

THE BEER

Thornbridge, Jaipur (5.9%), *IPA*, UK

If there's one beer that defines my career as a beer writer, it's this one. Not long after my second book was published in 2006, I got a phone call from Thornbridge, whom I'd never heard of before, asking me if I'd like to go and visit, with a view to doing an after dinner speech at an event they were holding.

I'd just come back from my first writing trip to America, where I'd tasted North American IPAs for the first time. I had no idea anyone was doing them in the UK at this point. I tasted Jaipur and screamed, 'This is it! This is that flavour!'

Since that point, Jaipur has gone on to win more awards than any other British beer over the last twenty years. While I haven't quite done that as a writer, my career grew in parallel with Thornbridge's success. We were both just starting out when we met, and we've both done OK.

Thornbridge effortlessly straddle the traditional British cask ale scene and being one of the original modern UK craft breweries. From best bitter to bold barrel-aged experiments, every beer they make is a classic example of its style. After twenty years, Jaipur remains their masterpiece, a beer I'll drink whenever I can.

THE SONG

New Order, 'Temptation', Factory Records, *Electronic Pop*

If there's one song that defines my passion as a music lover … well, actually there are two, maybe three now, after doing this book. But this was the first of them.

It was 1982 and I was thirteen years old. I was doing a paper round, so I'd been given a digital clock radio for Christmas so I could set my alarm for 6.30am and deliver papers before leaving for school. It had a 'sleep' function, which allowed me to set a timer for anywhere between one and 59 minutes for when I wanted the radio to switch off, before it came back on with the alarm at 6.30. For a few years, this meant I fell asleep listening to John Peel's late-night show on BBC Radio 1.

The tempo of music affects your heart rate when you sleep. For most people, it's beating at around 60bpm. Play music at the same tempo, and it helps regulate and relax your body. While the body slips into REM sleep, the brain continues to register the impact of ambient noise. Calming music is likely to inspire more positive dreams. The absolute limit you should listen to for nice, restful sleep is 80bpm.

'Temptation' clocks in at 120bpm. I was dreaming the first time it got into my head. It woke me up, the way the noise of a party next door wakes you up. But the party had already taken root in my subconscious. It was a glorious, euphoric noise. So I found out what it was, and then I played it at every party, every disco I went to as a teenager and beyond. It was there at my eighteenth, my twenty-first, and those of my closest friends. I danced to it at my wedding reception. The more special moments it soundtracked, the more it became a necessary part of the next one. It's been there at every significant, joyful moment in my life since my early teens, the soundtrack of every single happy memory.

THE PAIRING

This is obviously a very personal pairing for me, and the depth of it may not translate for everyone. In that way it's a throwback to my very first, unserious attempt to pair beers and tunes, when I joked that New Order pair with every great beer ever brewed.

But it does work. It's a principle I've used several times in this book, but an old-school IPA, with a malty backbone and decent bittering hops, gives you bitterness and sweetness to play with in a crossmodal pairing. I've done experiments earlier where different songs bring out the sweetness and then the bitterness. This, like The Blue Nile and Anspach & Hobday's IPA earlier, celebrates both at once – the sweetness in the synths, the bitterness in Hooky's driving, dominating bass line, all topped with the joy that runs through both the song and the beer.

Lost and Grounded Keller Pils with PJ Harvey 'Let England Shake'

THE BEER

Lost and Grounded, Keller Pils (4.8%), *Lager*, UK

After travelling and brewing around the world together for many years, Alex Troncoso and his partner, Annie Clements, decided to settle down in Bristol and start a brewery. The main reason they started their own brewery was so that Alex could brew a keller pils, and nudge what was once an obscure style outside Germany onto a wider stage.

Kellerbier was – is – a traditional Franconian beer style that evolved out of the constraints of lager brewing before refrigeration. Lagers were often brewed in caves to keep cool, conditioned for long periods. Kellerbier was lagered in barrels that had no bung in them, so the carbonation could escape. Only at the end of the lagering period was the barrel sealed for transport. The beer arrived at its destination fresh, unfiltered and unpasteurised. This could have meant it was served naturally hazy. But its closest English relative is cask ale, which it resembles in many ways. And cask ale is usually served clear.

Keller Pils is an adaption of tradition kellerbier, somewhere between that and a traditional German pilsner. It used German and Belgian pilsner malt and German hop varieties to create a soft yet hoppy unfiltered, hazy lager. In 2023, it was named best kellerbier in the world at the World Beer Cup. I'm not going to argue with that.

THE SONG

PJ Harvey, 'Let England Shake', Island, 2011, *Folk Rock*

PJ Harvey might just be the most important recording artist working today. That's different from saying she's my favourite person to listen to. (Although live, she's one of the best acts I've ever seen.) What I mean is, like David Bowie before her, she takes influences from poetry, visual art, film, literature – oh yes, and music – and weaves them into something new that potentially touches all points back in turn. You get the sense people will still be listening to and studying her a century from now.

PJ HARVEY

The 2011 album *Let England Shake* was named album of the year by pretty much every British music publication that conducted such polls at the time. Both the lyrics and the music are modern, contemporary – lazily, we'd use that awful term, 'alternative'.

But at the same time, they seem like the latest point on a continuum through England's history, one that's constantly looking back and forth along itself. Allusions to modern wars feel like they're part of a pattern that goes back for ever. Modern instruments simultaneously sound medieval.

The title rack – the opener on the album – tells us 'England's dancing days are done,' that 'the West is asleep', and England is 'weighted down with silent dead.' But this is not (quite) an anti-war protest song. It's more an observation from someone who sees through time and propaganda to truths as deep and permanent as bedrock.

THE PAIRING

This is a crossmodal pairing about dryness. There's something austere in the music, brittle and parched. At points it opens up and gives you a glimpse of something colourful, then turns away again.

The beer is similarly dry and crisp in flavour. That's the comparison. But in flavour pairings with beer and food, you can contrast as well as compare. While the similarities are there, the beer also serves as a counterpoint to the song. Lager, as a beer style, is defined in Britain as being about refreshment. This beer, in some ways, provides the liquid relief to the song's dryness.

There's also artistry in both: this is also a pairing about excellence. A brewer and an artist at the top of their game, their work consisting of different elements, all honed to perfection.

Schlenkerla Märzen and Chimay Blue with Claude Debussy 'Clair de Lune' and Jimi Hendrix 'All Along the Watchtower'

THE BEER (1)

Chimay, Blue (9%), *Dubbel,* Belgium

Among the famous Belgian Trappist Brews, Chimay is seen by some beer aficionados as the one that sold out. It got too big. Too … accessible. Worryingly, people who aren't even Belgian beer geeks might see it and buy it. They might even enjoy it.

I once found two 750ml bottles of Chimay Blue, covered in dust, at the back of my beer cellar. They were two years past their best before dates. I decided to use them in a carbonnade, but thought I'd just check them first, to see if they were more suitable as vinegar on chips.

They didn't end up as vinegar, or in the sauce for the carbonnade. I decanted them carefully, and served them alongside the meal, instead of red wine.

Drinking them was like swimming in layers of vinous sophistication. Caramel, chocolate and rich red berry fruit drifted past like shoals of exotic fish. Each sip revealed something new, before taking you back around a building symphony that just became ever more glorious.

THE BEER (2)

Schlenkerla, Märzen (4.6%), *Rauchbier,* Germany

About 500 years ago, maltsters in many parts of the world had no option but to dry their barley over fires made from wood or hay. They tried everything to get rid of the smoky taint this gave to the malt. When coke was invented – coal with all the noxious gases burnt out of it – it allowed malt to be roasted with a greater degree of control than ever before, from a material that imparted no undesirable flavours to it. Pale ale swiftly emerged, followed by clean pilsner lager. Modern beer was born.

But whenever technology advances, there's always someone who says, 'Well, maybe I liked the old ways.' Around Bamberg in Germany, speciality maltster Weyermann still produces smoked malts. On purpose. Some brewers, such as Schlenkerla, smoke their own.

Schlenkerla traces its origins back to a tavern first recorded in 1405, and the cellars survive from that time. The current brewery has been in operation here since 1678, and suggests without actually saying so that 'the original Schlenkerla smokebeer' has been brewed there ever since. The nose conjures smoked bacon and smoked cheese. But there's more than smoke here. There's also a big malt character giving aromas of chocolate and caramel. On the palate it's all about dark flavours: echoes of roasty Guinness, the sweetness of a brown ale, all rich and round. The smoke then returns as an aftertaste, bitter and wistful. If you think all lagers are the same, you have no idea.

THE SONG (1)

Claude Debussy, 'Clair de Lune', Third Movement of Suite Bergamasque, 1905, *Classical*

Claude Debussy was a French composer who had a massive influence on the development of twentieth-century classical music. 'Clair de Lune' ('moonlight') is inspired by the Symbolist Paul Verlaine's poem of the same name. The poem is a dazzling collection of imagery that evokes a moonlit dance, populated by 'sad and lovely' people who anticipate the more modern phrase 'dance like no one is watching'. It doesn't matter if the dance is crap. It doesn't matter that we are flawed and ungainly.

We're still beautiful in that moment. Or as a poetic, non-literal translation of the poem from the French would have it, 'soul sings in minor key of triumphant love and luck disbelieving its own song; its shadow cries with moonlight.'

Debussy is already transposing one artistic sense for another by taking that point about the 'soul sings in minor key', and turning those words into music. But there's another blurring of the senses here, particularly in the piece's middle section, which is described as a 'blue note'. More commonly associated with jazz or blues, blue notes are flatter or lower than the rest of the song would lead you to expect, and are often used to evoke emotion, usually sadness, hence 'the blues'. In 'Clair de Lune', the blue note is a low string section, or the lower of a piano keyboard, playing over a rippling accompaniment that suggests swimming in fountains.

Synaesthesia was not a recognised condition during Debussy's lifetime, but he clearly experienced it, and used to describe how he wrote music 'in blue'. He once wrote, 'I am more and more convinced that music is not, in essence, a thing which can be cast into a traditional and fixed form. It is made up of colours and rhythms.' I imagine him at his piano, big swathes of blue swirling through the air as he played.

THE SONG (2)

Jimi Hendrix, 'All Along the Watchtower', Reprise, 1968, *Rock*

It's claimed by Jimi Hendrix's manager, Chas Chandler, that the term 'heavy metal' comes from a *New York Times* review that described the sound of Hendrix's guitar as being like 'listening to heavy metal falling from the sky'. Nowhere does it sound more apocalyptic and massive than on this cover of a Bob Dylan song, widely considered to be the definitive version, among other covers by musicians such as U2, Pearl

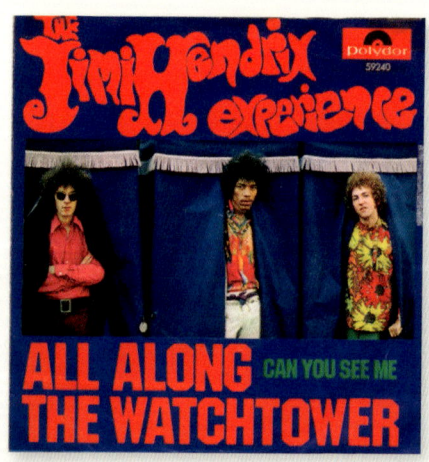

Jam, XTC and Bryan Ferry. The lyrics of the song recount a conversation between a 'joker' and a 'thief', with interpretations ranging from an allegory on Dylan's (the joker) relationship with his manager (the thief) to Biblical portents about the decline of civilisation.

THE PAIRING(S)

Both tracks feel like they belong with dark, heavy beers. Both the beers and the music are complex and multi-layered, swooping and diving around your consciousness. Each beer and each song could accurately by described as heavy, massive, brooding, and dense. But there are different kinds of darkness, and different kinds of complexity. Which beer goes best with which track and why? Opinion is often divided – but even though people disagree, they hold their respective views very strongly. Maybe try for yourself before reading on.

For me, both 'Clair de Lune' and Chimay evoke a sense of swimming in the dark, a mood of moonlit magical realism. The ripples of the blue note merge with the vinous elegance of the beer. Meanwhile, 'Watchtower' feels like it's crashing the party, stomping all over the delicacy of the beer and shattering the mood.

Likewise, Schlenkerla, does the same thing to 'Clair de Lune' – it's too astringent and aggressive, harshing the buzz. Turn things around though, and together, Schlenkerla and 'Watchtower' both feel like they belong in a heavily industrial landscape full of oil, steel and smoke. Equally aggressive, they nevertheless work together as powerful rather than threatening. The swirling of the guitar, the way it seems to own space, the size of it, is matched by the layers and depth of flavour in the beer. Chimay suddenly tastes watery, just as Debussy wilts under the smoke of Schlenkerla.

Orkney Brewery Dark Island Reserve with The Hope Blister 'Spider and I'

THE BEER

Orkney Brewery, Dark Island Reserve (10%), *Barrel-Aged Dark Ale*, UK

Orkney Dark Island is a dark ale, somewhere between a bitter and a porter, with roasted malts giving a fruity character with hints of coffee. Dark Island Reserve is a turbo-charged version of the beer that's aged for three months in old oak casks formerly used to mature fine scotch malt whisky.

Today, whisky-barrel-aged beers are relatively easy to find. They were rare when Orkney brewmaster Norman Sinclair decided he wanted to create a beer that made a real statement about not just his brewing skill, but also his home of Orkney, and beer itself, and where it was capable of going.

Pop the bottle open, and the beer pours oily, almost black, with very little carbonation. It's one of those rare beers where you become so absorbed in giving it a good sniff that you can carry on returning to it, forgetting to drink. The first time you go in you might get vanilla, tobacco or liquorice accenting the deep, dark, toasted malt. Go back again and it might be smoke or whisky, then figs and dates. The complexity is astonishing. With so much on the nose, it's not surprising that this variety of flavours develops on the palate, along with coffee, wood, spiciness, and alcohol warmth.

THE SONG

The Hope Blister, 'Spider and I', 4AD, 1998, *Ambient Noise*

If you read Side One of this book, you'll be familiar with the principle of music as organised noise. Well, here's some organised, exhilarating, slightly scary, magnificent noise for you. If you haven't heard this before, do yourself (and anyone near you) a favour. Plug in your headphones, turn them up as far as you can, and make sure you're sitting down. This is the aural equivalent of a deep tissue massage. You know it's doing you good, so you embrace the pain.

The Hope Blister were technically a reworking – or a sequel – of This Mortal Coil, who we've already heard from. But the artists working on Ivo Watts-Russell's collective project were completely different on this record, and the band name is different, so I'm allowing it.

'Spider and I' was the closing track on Brian Eno's 1977 album *Before and After Science*. It has a powerful synth riff that builds for the first ninety seconds of the song, suggesting something very powerful is about to come down the line. Then … it's nice, a bit whimsical, and it eventually fades out. It's half an idea, a sketch, really.

The Hope Blister are faithful to the original up to the point where the brief lyrics finish, almost as if Eno has thought of something else he'd rather be doing. Then, they basically say, 'Hi Brian, no offence, but we thought this song might sound … um … how do we say this, a little bit better if, at this point, after the lovely first bit, we might just, possibly … you know … um … UNLEASH HELL AND RAIN DOWN FIRE AND BRIMSTONE ON THE UNWORTHY SINNERS! AND YEA, THEY SHALL REPENT!'

It works pretty well.

THE PAIRING

The principle of this pairing is exactly the same as the one that guided The Neutrinos and Ampersand's Experiments in Evil earlier in the book. But it works so well, there's no harm in repeating it, just to show the principle really does work and that first one wasn't a fluke.

When you're pairing beer and food, one of the golden rules is to match equal levels of intensity. Pair a massive beer with a delicate starter and the beer will blow the food away. Similarly, here both the song and the beer are huge. They need to be paired on the principle of weight, volume and intensity. It's about turning both up to eleven and losing yourself to the explosion on your palate and the explosions in your ears, surrendering yourself to the crushing, glorious weight of both.

Bathams Best Bitter with The Real Sounds of Africa 'Wende Zako'

THE BEER

Bathams Best Bitter (4.3%), *Bitter,* UK

For about my first ten years in the job, whenever I told someone from the Midlands that I was a beer writer, their reaction was always the same.

'Have you tried Bathams?'

'No. Why does everyone keep asking me that?'

'Oh. Ah well. Never mind.'

It was a test I failed repeatedly. Until one day, the late food writer and beer fan Charles Campion asked me the same question. Instead of dismissing me as unworthy, he invited Liz and me to stay with him for a weekend. A few weeks later, we were in the Bull and Bladder – home of Bathams Brewery.

Charles ordered pints of Bathams Best. I'd been building up to this for years. It smelt like … bitter, the traditional earthy, spicy notes of Fuggles and Goldings hops on the nose. On the palate, it was a little too sweet for me, a bit watery perhaps? I went to take another sip to make sure, and was surprised that my glass was already empty. Surely only seconds had passed? But Charles was tapping his empty pint glass impatiently, saying, 'Your round, dear boy.'

Bathams pretends to be modest and unassuming. Somehow, it happens to be the most drinkable beer in the world.

THE SONG

The Real Sounds of Africa, 'Wende Zako', Cooking Vinyl, 1987, *Folk/World Music*

I first saw The Real Sounds of Africa at Glastonbury in 1987 – my first time there. My mates and I were listening to what was then dubbed 'world music', because it had become a staple on the late-night radio programmes we all tuned in to.

There was something a bit worthy and hippyish about the British evangelists of world music. But I loved the sounds that Zimbabwean bands made with guitars. The Bhundu Boys were the most famous. Because of them, others got exposure.

One night, John Peel played the Real Sounds' twelve-minute epic 'Tornados vs Dynamos', the story of a (real) football match that turned out to be a thrilling 3–3 encounter. Dynamos player Moses Chunga scored a hat-trick on his way to a record forty-six goals that season. 'Mose Chunga – it's a goal!' was our catch phrase for the year.

We made sure we got to the front so we could chant along with the Tornados v Dynamos commentary, especially for that all-important hat-trick. And then they finished with Wende Zako. Unlike the football song, it wasn't sung in English, apart from the simple, joyful chorus of 'Yeah, yeah yeah, mama, yeah yeah yeah!' We were singing it for the rest of the weekend, the guitar and horns under our skins, not letting go.

THE PAIRING

There are more pairings about bitterness and sweetness in this book than anything else, because bitterness and sweetness are the defining characteristics of beer flavour.

If this book contributes to the academic understanding of the crossmodal relationships between sound and flavour in any way, I hope it's that we challenge the perception that bitterness and brass go together on the basis that both are unpleasant. We disproved this already with Barnsley Bitter and Grimethorpe Colliery Band. This is a more emphatic example. This beer is sweeter than most best bitters are. This sweetness corresponds with the vocal harmonies and tinkling guitars that define Zimbabwean pop. But there's also bitterness, and the band also have a brass section that alternates with those sweet vocals.

It turns out that, after all this, English best bitter and Zimbabwean pop is one of the best beer and music pairings you can ever do.

Little Pomona Old Man & The Bee with Underworld 'Rez'

THE CIDER

Little Pomona, Old Man & The Bee (7.4%), *Cider*, UK

I began writing about cider because people asked me to. I'm a beer writer, and beer and cider are the same thing, aren't they? No, they're not. Cider is closer to wine than beer, but also very much a great drink in its own right that owes nothing to either beer or wine.

Over the past ten years or so, a new generation of cider makers has emerged with the mission of repositioning cider in our minds and undoing some of the damage done to its reputation by large commercial brands. Little Pomona is one of the very best of them. Founded by James and Susanna Forbes in 2017, they make ciders that showcase specific apple varieties in the way winemakers do grapes. These drinks are about a sense of place, of care and love.

Apples change from year to year, dependent on weather and harvest. Old Man & The Bee is the closest Little Pomona get to a core product. It's made every year from their fruit of their own orchard, a little different each time. As I write, the 2020 vintage is still on sale. There's a lot of fruit and spice here, and some prominent acidity. The former white burgundy barrels in which the cider spent eleven months give more spice and a bit of vanilla on top.

THE SONG

Underworld, 'Rez', Smith Hyde Productions/ Universal, 1993, *Progressive Trance*

I have a weird relationship with dancing. Weird in that, I'm so bad at it, I'm terrified of it. Remember when you pulled funny faces as a kid and your parents warned that if the wind changed, you'd stay like that? Similar things can happen in reality.

In my twenties, I used to do an ironic, idiot dad-dance to make people laugh. The wind changed, and it stuck, becoming my actual dance. The only one I can do.

I love Underworld because they make it so that this doesn't matter. I can actually dance to their music in my mind, while sitting at my desk.

In some form, Underworld go back a very long way. But in their current incarnation, they've been making progressive dance music since the early 1990s. By bringing in elements of pop, rock and beat poetry over the complex and driving beats, they crossed over from dance to mainstream with huge success.

'Rez' is unusual in that it's an instrumental, lacking vocalist Karl Hyde's hypnotic, shamanistic lyrics. Live, this leaves Hyde free to dance beneath a single overhead spot, with movements that would have had him drowned as a witch or worshipped as the god Pan, depending on which past century he did them in. This dance is the one I do in my brain when I hear this song. The dance no one else can see.

THE PAIRING

As we've established, cider is not beer. That's the point we're making with this pairing. This song needs a cider, not a beer.

When I was writing *World's Best Cider* with photographer Bill Bradshaw, I created a soundtrack for the hours we were going to be spending driving through the countryside of the UK and the United States. I played it to Liz the night before we left. Every now and again, she'd say, 'That's not cider, that's more wine.' Or 'That one's definitely beer.' When I stripped these songs out, I was left with light, airy folk and alt.country. Songs that sounded like the outdoors and the countryside to a city-based couple.

But there was also 'Rez'. It slammed the playlist shut at the end. Every time I hear it, it sounds like the last track on the last night of every music festival I've been to. And at these festivals, when you're out in the fields and hills, there's no better drink than strong, artisanal cider. In a crossmodal sense, the characteristic acidity of Little Pomona corresponds perfectly with the delirious dissonance in the melody.

Endnotes

OVERTURE (*pages 11 to 13*)

1. Okay, not dancing. Liz doesn't let me dance. It gives her the fear.

2. It's on page 68 and it's a killer. Twelve years later I have yet to write another joke that's as funny.

3. Paired with the wonderful banger 'Holiday' by Confidence Man.

Side One

TRACK 1 (*pages 17 to 36*)

4. Not as much fun as it sounds.

5. Source: 'You ask the questions': Jilly Goolden, *The Independent*, 24 November 1999.

6. This was shortly after the agency in question had been told to make an ad to try to persuade people of the implausible notion that Kellogg's Bran Flakes tasted nice. The response was a jingle that went, 'They're tasty, tasty, very, very tasty. They're very tasty'. That was honestly the best they could do.

7. Source: Manolaraki, Eleni Hall, 'Senses and the Sacred in Pliny's Natural History', in Illinois Classical Studies, vol 43, no 1, pp.207–233 University of Illinois Press, Spring 2018.

8. Source: Danijela Kambaskovic-Sawers, Charles T. Wolfe: 'The Senses in Philosophy and Science: From the Nobility of Sight to the Materialism of Touch'. In *A Cultural History of the Senses in the Renaissance*, https://hal.science/hal-02069998v1, 2014.

9. Source: Majid, A, Roberts, SG, Cilissen, L, Levinson, SC, 'Differential coding of perception in the world's languages', PNAS, vol. 115 no 45, November 2018.

10. Source: Kant, I. *Kant: Anthropology From a Pragmatic Point of View*, Cambridge University Press, 2006, [1798].

11. Source: Spence, C, Smith, B, Auvray, M, 'Confusing tastes with flavours', in *Perception and Its Modalities*, Oxford University Press, 2014.

12. Source: Spence et al, *ibid.*

13. It's also why beer is a better pairing with cheese than wine. The carbonation lifts the fat of the cheese off your palate, releasing more flavour compounds and refreshing the taste buds, so you get a lot more from the cheese.

14. Source: Breslin Paul A S, 'An Evolutionary Perspective on Food and Human Taste', *Current Biology* 23, R409–R418, 2013, Elsevier Ltd.

15. Source: Brillat-Savarin, Jean-Anthelme, *The Physiology of Taste, Or Meditations on Transcendental Gastronomy* (1826), Knopf Doubleday Publishing Group; Reprint edition (4 October 2011).

TRACK 2 (*pages 45 to 56*)

16. Source: Oxenham, Andrew, 'How We Hear: The Perception and Neural Coding of Sound', in *Annual Review of Psychology*, 2019.

17. Source: Beament, Sir James, *How We Hear Music: The Relationship Between Music and the Hearing Mechanism*, Boydell Press, Suffolk, 2003.

18. I wondered if the rumour was true that Michael Jackson's song 'Ben' was about his pet rat, or whether that was myth. The truth is far more bizarre. 'Ben' was originally written by Don Black and Walter Scharf for the 1972 movie of the same name. The film Ben was a sequel to a film called Willard. Ben is Willard's pet rat. Willard can control Ben and other rats, and gets them to kill for him. At the end, the rats turn against him and kill him. In the sequel, Danny, a young boy with a serious heart condition, finds Ben in a shed, and they become best mates. But Ben turns into an evil super rat, head of a gang of rats that kill humans to get to food. But not Danny, who has written a song about how much he loves Ben. The rats decide he's alright.

TRACK 3 (*pages 60 to 76*)

19. Source: Levitan C A, Ren, J, Woods, A T, Boesveldt S, Chan JS, McKenzie KJ, Dodson, M, Levin JA, Leong, CXR, van den Bosch, JJF 'Cross-Cultural Color-Odor Associations', PLOS ONE vol issue 7, July 2014.

20. Brown, Pete, *Three Sheets to the Wind: One Man's Quest for the Meaning of Beer*, Pan Macmillan, London, 2006.

21. Obama had a Bud Light, Gates a Sam Adams Light, and Buckler a Blue Moon.

22. Source: Zappa, Frank, with Occhiogrosso, Peter, *The Real Frank Zappa Book*, Simon & Schuster, New York, 1989.

23. See page 116.

24. If you ever read this, unlikely as that may be – soz, Carolyn.

25. This is a lot funnier with the pictures.

26. Source: Limb, CJ, Braun AR, 'Neural Substrates of Spontaneous Musical Performance: An fMRI Study of Jazz Improvisation', National Library of Medicine (US), 2008.

27. And there I am, being uncomfortable and apologetic about my ability to describe and evoke flavour, because, even now, I'm still acutely self-conscious that this is something most of us feel uncomfortable about.

28. Source: Prescott, John, *Taste Matters – Why We Like the Foods We Do*, Reaktion Books, London, 2012.

29. Source: Spence, Charles, *The Perfect Meal – The Multisensory Science of Food and Dining*, Wiley & Sons, Chichester, 2014.

30. Source: Spence, *ibid*.

31. Source: North, AC, Hargreaves, DJ, McKendrick, J, 'The influence of In-store Music on Wine Selections', in *Journal of Applied Psychology* no 84, 1999.

32. Source: Byrne, David, *How Music Works*, Canongate, Edinburgh, 2012.

TRACK 4   *(pages 83 to 99)*

33. These experiments are summarised in Spence, Charles, 'Crossmodal Correspondences: A Tutorial Review', Psychonomic Society Inc., 2011.

34. Source: Spence, *ibid*.

35. Source: Köhler, W, *Gestalt Psychology*, Liveright, New York, 1929.

36. Source: Sidhu, DM, and Vigliocco, G, 'I Don't See What You're Saying – The Maluma/Takete Effect Does Not Depend on the Visual Appearance of Phonemes As They Are Articulated', *Psychonomic Bulletin and Review*, 2022.

37. Source: https://www.bbc.co.uk/news/business-26925249

38. Source: Holt-Hansen, K, 'Taste and Pitch', *Perceptual and Motor Skills*, 27(1), 1968.

39. Source Holt-Hansen, K, 'Extraordinary Experiences During Cross-Modal Perception', *Perceptual and Motor Skills*, 43(3), 1976.

40. Source: Crisinel, A-S, and Spence, C, 'A Fruity Note: Crossmodal Associations Between Odors and Musical Notes', *Chemical Senses*, 37, 2011.

41. Source: Carvalho, FR, Wang, Q, De Causmaecker, B, Steenhaut, K, Van Ee, R, Spence, C, 'Tune That Beer! Listening for the Pitch of Beer', *Beverages*, 2016.

42. Source: Crisinel, A-S, and Spence, C, As Bitter as a Trombone: Synesthetic Correspondences in Nonsynesthetes Between Tastes/Flavors and Musical Notes, *Attention, Perception, & Psychophysics*, 72 (7), 2010.

43. I don't always do this, but my data is based on about 450 responses. While my 'experiments' were carried out in rooms above pubs, small concert venues and tents at festivals, with people drinking beer, my sample size is far bigger than that of any of the real experiments discussed here, which has to count for something.

44. Source: Carvalho, FR, Wang, Q, De Causmaecker, B, Steenhaut, K, Van Ee, R, Spence, C, 'A Bittersweet Symphony: Systematically Modulating the Taste of Food by Changing the Sonic Properties of the Soundtrack Playing in the Background', *Food Quality and Preference* 24, 2012.

45. Source: Carvalho, FR, Wang, Q, Van Ee, R, Spence, C, 'The Influence of Soundscapes on the Perception and Evaluation of Beers', *Food Quality and Preference*, 52, 2016.

46. Which I slightly take issue with. I'd been exploring 'the influence of sound-scapes on the perception and evaluation of beers' for four years by this point, but my stuff doesn't count as 'proper' research. To be fair, an article I'd written for *Word* magazine did get a mention.

Further reading

Books

ACKERMAN, Diane, *A Natural History of the Senses*, Orion, London, 1990.

BEAMENT, Sir James, *How We Hear Music: The Relationship Between Music and the Hearing Mechanism*, Boydell Press, Suffolk, 2003.

BRILLAT-SAVARIN, Jean-Anthelme, *The Physiology of Taste, Or Meditations on Transcendental Gastronomy*, (1826), Knopf Doubleday Publishing Group; Reprint edition (4 October 2011).

BROWN, Pete, *Three Sheets to the Wind: One Man's Quest for the Meaning of Beer*, Pan Macmillan, London, 2006.

BYRNE, David, *How Music Works*, Canongate, Edinburgh, 2012.

HOLMES, Bob, *Flavour: A User's gGide to our Most Neglected Sense*, Penguin, London, 2017.

KANT, I, *Anthropology From a Pragmatic Point of View,* Cambridge University Press. 2006, [1798].

KÖHLER, W, *Gestalt Psychology*, Liveright, New York, 1929.

KORSMEYER, Carolyn, *Making Sense of Taste: Food and Philosophy*, Cornell University Press, New York, 1999.

LEVITIN, Daniel, *This is Your Brain on Music: Understanding a Human Obsession,* Atlantic Books, London, 2006.

LUCE, R Duncan, *Sound & Hearing: A Conceptual Introduction*, Psychology Press, New York, 1993.

MCQUAID, John, *Tasty: The Art and Science of What We Eat,* Scribner, New York, 2015.

PRESCOTT, John, *Taste Matters – Why We Like the Foods We Do*, Reaktion Books, London, 2012.

ROGERS, Jude, *The Sound of Being Human: How Music Shapes our Lives*, White Rabbit, London, 2022.

ROSENBLUM, Lawrence D, *See What I'm Saying: The Extraordinary Power of our Five Senses,* WW Norton & Co, New York, 2010.

SHEPHERD, Gordon, *Neurogastronomy: How the Brain Creates Flavour and Why it Matters*, Columbia University Press, New York, 2012.

SPENCE, Charles, PIQUERAS-FISZMAN, Betina, *The Perfect Meal – The Multisensory Science of Food and Dining,* Wiley & Sons, Chichester, 2014.

SPENCE, Charles, *Gastrophysics: The New Science of Eating*, Penguin, London, 2018.

STEIN, Barry, & MEREDITH, M Alex, *The Merging of the Senses*, Massachusetts Institute of Technology, Boston, 1993.

STUCKEY, Barb, *Taste: Surprising Stories and Science About Why Food Tastes Good*, Simon & Schuster, New York, 2012.

THIS, Hervé, *Molecular Gastronomy: Exploring the Science of Flavour*, Columbia University Press, New York, 2002.

WILLIAMSON, Victoria, *You Are the Music: How Music Reveals What it Means to be Human*, Icon, London, 2014.

ZAPPA, Frank, with OCCHIOGROSSO, Peter, *The Real Frank Zappa Book*, Simon & Schuster, New York, 1989.

Academic papers

MALIKA, Auvray, M, & SPENCE, C, 'The Multisensory Perception of Flavor', *Conscoiusness and Cognition*, 2007.

BRESLIN, Paul, AS, 'An Evolutionary Perspective on Food and Human Taste', *Current Biology* 23, R409–R418, 2013, Elsevier Ltd.

BURZYSNKA, Jo, 'Wine and Music: The Synergies Between Sound and Taste', International Food Design Conference and Symposium, New Zealand, July 2014.

BRONNER, K, FRIELER, K, BRUHN, H, HIRT, R, & PIPER, D, 'What is the Sound of Citrus? Research on the Correspondences between the Perception of Sound and Flavour', Proceedings of the ICMPC, 2012.

CARVALHO, FR, WANG, Q, DE CAUSMAECKER, B, STEENHAUT, K, VAN EE, R, & SPENCE, C, 'A Bittersweet Symphony: Systematically Modulating the Taste of Food by Changing the Sonic Properties of the Soundtrack Playing in the Background', *Food Quality and Preference* 24, 2012.

CARVALHO, FR, WANG, Q, VAN EE, R, & SPENCE, C, 'The Influence of Soundscapes on the Perception and Evaluation of Beers', *Food Quality and Preference* 52, 2016

CARVALHO, FR, WANG, Q, DE CAUSMAECKER, B, STEENHAUT, K, VAN EE, R, & SPENCE, C, 'Tune That Beer! Listening for the Pitch of Beer', *Beverages*, 2016.

CRISINEL, A-S, & SPENCE, C, 'Implicit Association Between Basic Tastes and Pitch', *Neuroscience Letters*, 2009.

CRISINEL, A-S, & SPENCE, C, 'As Bitter as a Trombone: Synesthetic Correspondences in Nonsynesthetes Between Tastes/Flavors and Musical Notes', *Attention, Perception, & Psychophysics*, 72 (7), 2010.

CRISINEL, A-S, & SPENCE, C, 'A Fruity Note: Crossmodal Associations Between Odors and Musical Notes', *Chemical Senses*, 37, 2011.

CRISINEL, A-S, JONES, S, & SPENCE, C, 'The Sweet Taste of Maluma: Crossmodal Associations Between Tastes and Words', *Chemosensory Perception*, 2012.

CRISINEL, A-S, COSSER, S, KING, S, JONES, R, PETRIE, J, & SPENCE, C, 'A Bittersweet Symphony: Systematically Modulating the Taste of Food by Changing the Sonic Properties of the Soundtrack Playing in the Background', *Food Quality and Preference*, 2011.

GUEDES, D, van GARRIDO, M, LAMY, E, & Cavalheiro, BP, 'Crossmodal Interactions between Audition and Taste: A Systematic Review and Narrative Synthesis', *Food Quality and Preference*, 2022.

GUETTA, R, & LOUI, P, 'When Music is Salty: The Crossmodal Associations between Sound and Taste', *PLOS One*, 2017.

KAMBASKOVIC-SAWERS, D, & WOLFE, CT, 'The Senses in Philosophy and Science: From the Nobility of Sight to the Materialism of Touch': *A Cultural History of the Senses in the Renaissance*, https://hal.science/hal-02069998v1, 2014.

KNÖFERLE, K, & SPENCE, C, 'Crossmodal Correspondences between Sounds and Tastes', *Psychonomic Bulletin and Review*, 2012.

HOLT-HANSEN, K, 'Taste and Pitch', *Perceptual and Motor Skills*, 27(1), 1968.

HOLT-HANSEN, K, 'Extraordinary Experiences During Cross-Modal Perception', *Perceptual and Motor Skills*, 43(3), 1976.

LEVITAN, CA, REN, J, WOODS, AT, BOESVELDT, S, CHAN, JS, McKENZIE KJ, DODSON, M, LEVIN, JA, LEONG, CXR, & van den BOSCH, JJF, 'Cross-Cultural Color-Odor Associations', *PLOS One*, 7, July 2014.

LIMB, CJ, & BRAUN, AR, 'Neural Substrates of Spontaneous Musical Performance: An fMRI Study of Jazz Improvisation', National Library of Medicine (US), 2008.

MAJID, A, ROBERTS, SG, CILISSEN, L, & LEVINSON, SC, 'Differential Coding of Perception in the World's Languages', *PNAS*, vol 115, no 45, November 2018.

MANOLARAKI, Eleni Hall, 'Senses and the Sacred in Pliny's Natural History,' in *Illinois Classical Studies*, vol 43, no 1, University of Illinois Press, 2018.

MESZ, B, SIGMAN, M, & TREVISAN, MA, 'The Taste of Music', *Perception*, 2011.

MESZ, B, SIGMAN, M, & TREVISAN, MA, 'A Composition Algorithm Based on Crossmodal Taste-Music Correspondences', *Frontiers in Human Neuroscience*, 2012.

MESZ, B, HERZOG, K, AMUSATEGUI, JC, SAMARUGA, L, & TEDESCO, S, 'Let's Drink This Song Together: Interactive Taste-Sound Systems', Proceedings of 2nd ACM SIGCHI International Workshop on Multisensory Approaches to Human-Food Interaction, New York, 2017.

NORTH, AC, HARGREAVES, DJ, & McKENDRICK, J, 'The Influence of In-store Music on Wine Selections', *Journal of Applied Psychology*, 84, 1999.

NORTH, Adrian C, 'The Effect of Background Music on the Taste of Wine', *British Journal of Psychology*, 103, 2012.

OXENHAM, Andrew, 'How We Hear: The Perception and Neural Coding of Sound', *Annual Review of Psychology*, 2019.

PRESCOTT, John, 'Multisensory Processes in Flavour Perception and their Influence on Food Choice', *Current Opinion in Food Science*, 3, 2015.

SHANG, Nan, & STYLES, Suzy J, 'Is a High Tone Pointy? Speakers of Different Languages Match Mandarin Chinese Tones to Visual Shapes Differently', *Frontiers in Psychology*, 8, 2017.

SIDHU, DM, and VIGLIOCCO, G, 'I Don't See What You're Saying – The Maluma/Takete Effect Does Not Depend on the Visual Appearance of Phonemes as they are Articulated', *Psychonomic Bulletin and Review*, 2022.

SPENCE, Charles, 'Crossmodal Correspondences: A Tutorial Review', *Psychonomic Society Inc*, 2011.

SPENCE, Charles, & DEROY, Ophelia, 'How Automatic are Crossmodal Correspondences?' *Consciousness and Cognition*, 12, 2013.

SPENCE, C, SMITH, B, AUVRAY, M, 'Confusing Tastes with Flavours', *Perception and Its Modalities*, Oxford University Press, 2014.

VAN BEILEN, M, BULT, H, RENKEN, R, STIEGER, M, THUMFART, S, CORNELISSEN, F, KOOIJMAN, V, 'Effects of Visual Priming on Taste-Odor Interaction', *PLOS One*, 9, 2011.

JUSTUS V VERHAGEN, JV, & ENGELEN, L, 'The Neurocognitive Bases of Human Multimodal Food Perception: Sensory Integration', *Neuroscience and Behavioural Reviews*, 2005.

WOODS, AT, POLIAKOFF, E, LLOYD, DM, KUENZEL, J, HODSON, R, GONDA, H, BATCHELOR, J, DIJKSTERHUIS, GB, & THOMAS, A, 'Effect of Background Noise on Food Perception', *Food Quality and Preference*, 2010.

ZAMPINI, M, SANABRIA, D, PHILLIPS, N, & SPENCE, C, 'The Multisensory Perception of Flavor: Assessing the Influence of Color Cues on Flavor Discrimination Responses', *Food Quality and Preference*, 18, 2007.

Mainstream media articles

'You Ask the Questions', Jilly Goolden, *The Independent*, 24 November 1999, independent.co.uk/news/people/profiles/you-ask-the-questions-jilly-goolden-743139.html

'Sweet or Sour? Altering how we Taste our Food', BBC, 16 April 2014, bbc.co.uk/news/business-26925249